ADVANCED ADAPTIVE CONTROL

Related Pergamon Titles

Books

BANYASZ
Adaptive Systems

BOULLART *et al.*
Application of Artificial Intelligence to Process Control

DEVANATHAN
Intelligent Tuning and Adaptive Control

DUGARD *et al.*
Adaptive Systems in Control and Signal Processing

LOTOTSKY
Evaluation of Adaptive Control Strategies

NAJIM and POZNYAK
Learning Automata: Theory and Applications

SINGH
Advances in Systems, Control and Information Engineering

SINGH
Systems and Control Encyclopedia

VENTRIGLIA
Neural Modelling and Neural Networks

Journals

Automatica
Chemical Engineering Science
Computers and Structures
Control Engineering Practice
Engineering Applications of Artificial Intelligence
Neural Networks
Transportation Research

Full details of all Elsevier Science Journals are available on request
from your nearest Elsevier Science office

ADVANCED ADAPTIVE CONTROL

by

H. WANG

Department of Paper Science, UMIST, U.K.

G. P. LIU

Department of Automatic Control and Systems Engineering,
Sheffield University, U.K.

C. J. HARRIS and M. BROWN

Department of Electronics and Computer Science,
University of Southampton, U.K.

PERGAMON

U.K.   Elsevier Science Ltd, The Boulevard, Langford Lane, Kidlington,
            Oxford, OX5 1GB, U.K.

U.S.A.   Elsevier Science Inc., 660 White Plains Road, Tarrytown,
            New York 10591-5153, U.S.A.

JAPAN   Elsevier Science Japan, Tsunashima Building Annex, 3-20-12 Yushima,
            Bunkyo-ku, Tokyo 113, Japan

First edition 1995

Library of Congress Cataloging in Publication Data

Advanced adaptive control / H. Wang ... [et al.].
p. cm.
Includes index.
1. Adaptive control systems. I. Wang, H. (Hong)
TJ217.A3857 1995
629.8'36-dc20   95-11906

British Library Cataloguing in Publication Data

A catalogue record for this book is available from the British
Library

ISBN 0 08 0420206

Printed in Great Britain by Redwood Books, Trowbridge, Wiltshire

Contents

Preface

Adaptive control has been a rich area for basic theoretical research into the autonomous control of *a priori* unknown dynamical processes. Much of the three decades of endeavour has been naturally devoted to studies and applications associated with linear time invariant processes subject to Gaussian disturbances or mismodelling errors. Extending that theory to encompass temporal and spatial parametric variations (through operating point changes), nonlinear dynamics, and non-Gaussian disturbances/distributions is within the province of *Advanced Adaptive Control*, aspects of which are addressed in this book.

Increasing the generality and complexity of adaptive control theory inevitably leads to a decreasing rate of progress via conventional mathematical methods (such as nonlinear time series analysis, frequency domain methods), and progress can only be made by evolving or searching for different methods of systems analysis representation and synthesis. In this latter context *intelligent control* methods based upon ideas and techniques from the disciplines of neurophysiology, cognitive sciences, operational research, approximation theory and control theory appear to open new research opportunities in adaptive control. This book addresses some of the major issues and methods in *Advanced Adaptive Control* via some recent results of the authors - specifically we consider the possibility of:-

- Utilising adaptive or self-organising artificial neural networks, fuzzy logic and rule based methods to solve nonlinear adaptive control problems for unknown plants.

- Extending the current self-tuning control strategies to *a priori* unknown linear systems subject to general external disturbances.

- Constructing adaptive control schemes for *singular* systems, for which the system is expressed as a combination of dynamic and algebraic equations.

In considering these aspects of advanced adaptive control, it is assumed that the reader is familiar with linear systems theory, classical control theory, linear parametric identification theory and some elementary knowledge of adaptive control, found in the excellent texts of Goodwin and Sin (1984), Navendra and Annaswamy (1989), Astrom and Wittenmark (1989) and Wellstead and Zarrop (1991). Much of the work described in this book is based upon a series of publications by the authors, and the following publishers are gratefully acknowledged for permission to publish aspects of our work that appeared in their journals: The Institute of Electrical Engineers, The Institute of Measurement and Control, Kluwer Academic Publishers, Elsevier Science Publishers (BV) and Taylor and Francis Ltd. Also the authors acknowledge their various departments, colleagues and research students for valuable comments, suggestions and criticisms during this research and writing. Particular thanks goes to Miss E. Harris and Mrs J. Wood of UMIST for their skills and diligence in preparing the manuscript. Also, thanks to EPSRC and Lucas Aerospace for their generous financial support for some of this research. And finally, a special thanks to our wives, Li Mei, Wei Hong, Joy and Adrienne for their constant encouragement, good humour and patience.

Hong Wang

Guo Ping Liu

Chris John Harris

Martin Brown

February 1995

Chapter 1

Introduction

1.1 Introduction

Adaptive control and its application to a wide range of industrial and commercial processes has been the subject of intensive investigation for over three decades. Since the early 1960's there has been an exponential growth in adaptive control publications, the earliest publications originated in aerospace for flight control and continues to this day to attract attention for high incidence, high performing aircraft. The need for adaptive control remains that of overcoming the difficulty of determining suitable control laws that are satisfactory over a wide range of operating points. Alternate methods such as H_∞ control solve process parameter variations (due to temporal, spatial or operating point changes) by incorporating *known* parameter variations into a robust off-line design procedure, inevitably the resultant controller trades robustness or parameter insensitivity for performance, usually in the form of a high order, conservative controller. If the dynamic process is highly nonlinear or with unknown dynamics and associated parametric variations, H_∞ methods are inadequate; although current H_∞ research is focused on nonlinear multivariable processes [45]. Fixed controllers such as H_∞, PID, model reference, LQG/LQR, pole placement and feedback linearisation are determined off-line via open loop *a priori* plant knowledge in a performance independent fasion. The various adaptive control schemes discussed in this book are *performance based* utilising varying amounts of past information. Modern adaptive control such as intelligent control utilises almost entirely past experimential knowledge.

Adaptive control uses both *a priori* knowledge about the controlled process, as well as the automatic incorporation of acquired outline knowledge based on observations of the process. Adaptive control is an extension and generalisation of classical off-line feedback control synthesis, by which all or some of the controller parameters are adjusted automatically in response to online process observations or output residuals (hence the name self-tuners for a subset of adaptive controllers). Adaptive control which utilises only feedforward disturbance measurements is a form of gain scheduling, this popular and effective methodology relies critically on accurate *a priori* plant knowledge and knowledge of the online feedforward measurements (such as the operating point). Adaptation of the controller in response to feedback measurements from the *a priori* unknown controlled process is central to adaptive control research and is the objective of this book.

Adaptive control uses performance feedback to adjust controller parameters and structure online. Conventional adaptive control such as model reference adaptive control, and self-tuning have significant problems with unmodelled nonlinearities. Whereas the more recent research into learning adaptive control based on artificial neural networks and neurofuzzy algorithms address system unknown nonlinearities directly by functional approximation, utilising significant amounts of memory or past information and performance feedback to adjust the controller structure and associated parameter outline.

Online learning for nonlinear processes introduces another set of problems such as guaranteed learning convergence and closed loop stability conditions, and the limitation of methods with provable learning conditions to low input dimensional problems (the so called Gabor-Kolmogorov curse of dimensionality).

Depending on the nature of the plant (assumed unknown) in adaptive control, different adaptive control strategies may be applied. There are two general approaches - *direct adaptive control* whereby the controller is synthesised without the construction of a process model, utilising performance based information only, and *indirect adaptive control* whereby a process/plant model is constructed (usually via online observation) which in turn is utilised to generate an appropriate controller. The indirect controller whilst apparently more complex does allow some form of separation between the model estimation and controller determination (the certainty-equivalence principle) - ensuring individual optimisation and selection. Also a wide variety of adaption schemes exist within these two generic classes, such as Model Reference Adaptive Control (MRAC; Landau 1979, [83]), Self-Tuning Adaptive Control (STAC; Harris and Billings 1981, [63]), Gain Scheduling (GS; Astrom and Wittenmark, 1989, [12], Lawrence and Harris, 1993, [85]), Self Organising Fuzzy Logic (SOFLIC; Harris, Moore, Brown 1993), Artificial Neural Networks (ANN; Warwick et al 1992) and Neurofuzzy Adaptive Control (NAC; Brown and Harris 1994, [19]).

1.2   Adaptive Control Schemes

1.2.1   Adaptive control for linear systems: MRAC

Research on adaptive control was mainly focused on unknown linear systems during 1980's. The system model was expressed either by a set of linear differential equations or by a set of difference equations, whose coefficients are unknown. Many control strategies were developed, and among them MRACs and STACs are the most representative. Figure 1.1 shows the basic structure of an MRAC system [82, 115].

It can be seen that the system consists of the following four components:

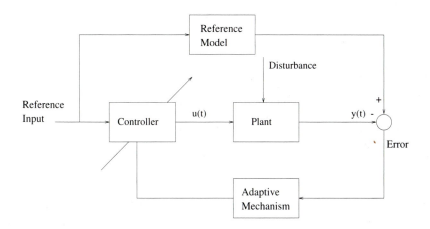

Figure 1.1: Model reference adaptive control system.

i) The plant represented by a transfer function model with unknown parameters;

ii) A reference model which generates a desired closed loop system output response;

iii) A controller with time-varying parameters

iv) An adaptive mechanism which makes use of the differences between the reference model and controlled plant to adjust the time varying parameters of the controller.

The purpose of the controller and adaptive mechanism design is to stablize the closed loop system and to ensure that the output of the system follows that of the reference model as closely as possible. Of course, desired dynamic behaviour, such as fast and accurate response, should also be achieved. Since stability of the closed loop is very important, Lyapunov design [125] is widely used for MRAC systems, where a Lyapunov function is constructed and the adaptive mechanism is obtained such that the derivative of the Lyapunov function is negative to ensure closed loop system stability.

The first MARC system was developed at MIT in 1950's and was further studied by Moor and Coop in 1962. The uniform structure was proposed by Landau in 1979 [83] and different realizable forms were investigated by Narendra, Kreissmmeler and many other researchers in later 70's, [116, 117, 81, 47]. In 1980, the remarkable result on closed loop stability (uniform boundness of all the variables within the closed loop system) for input-output based MARC system was established by Narendra and his colleagues [116, 117]. Subsequently, research has mainly concentrated on the improvement of robustness.

1.2.2   Self-tuning adaptive control systems

Compared with MRAC systems, most STAC systems only have three components; the plant to be controlled, the adaptive mechanism and the controller. In this case the adaptive mechanism is composed by a system identification unit and an on-line control parameter evaluation unit. The identification uses the information of the plant input and output to estimate the unknown plant parameters. The estimated plant parameters are then used to evaluate the parameters of the controller. Figure 1.2 shows the general structure of self-tuning adaptive control systems.

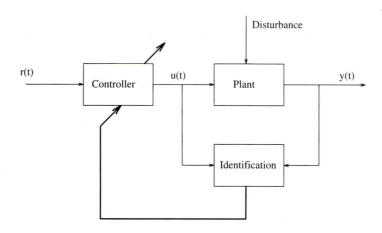

Figure 1.2: Self-tuning adaptive control systems.

Since system identification and controller parameter evaluation can be performed separately, different identification algorithms [104] can be combined with different controllers to produce different self-tuning strategies. Minimum variance self-tuning control [12], d-step-ahead predictive control [57] and pole-zero assignment self-tuning control [164] are examples of these combinations. Therefore, STAC system is more flexible and is more convenient to use than MRAC. Again, the purpose of the STAC system design is to stabilize the closed loop system and realize desired tracking with respect to demand signals. After self-tuning adaptive control was proposed by Astrom and Wittenmark in 1973 [8], investigation into both control structure and identification algorithms has long been a fascinating field in adaptive control for linear systems. Identification algorithms leading to fast convergent estimation and the generalisation of the desired dynamic performance were derived. Moreover, it has been shown that STAC and MRAC systems are equivalent under some certain conditions, [47].

1.2.3 Adaptive control for nonlinear systems

There remain many unsolved problems in nonlinear systems control, in particular, the design and implementation of adaptive control for nonlinear process is extremely difficult. In most cases the control strategies developed largely depend on the particular information on the nonlinear structure of the plant. The nonlinear adaptive control for robot arms is one such example. Therefore, unified or relatively unified adaptive control methods, such as MRAC and STAC for unknown linear systems, still need extension to nonlinear unknown systems.

1.2.4 Rule-based adaptive control

The adaptive control strategies discussed in sections 1.2.1-1.2.3 rely on the mathematical model (with unknown coefficients) of the plant, which is difficult to obtain in some cases. Rule-based adaptive tuning and control can therefore be used as an alternative [151]. Rule-based tuning normally consists of a set of rules and a rule selector (reinforcer). The selector monitors the performance and the state of the system and decides which rule should be used. Rule-based tuning becomes adaptive when the rule set needs to be updated in order to cope with changes and uncertainties of the plant. The structure of rule-based adaptive tuning control system is shown in Fig. 1.3.

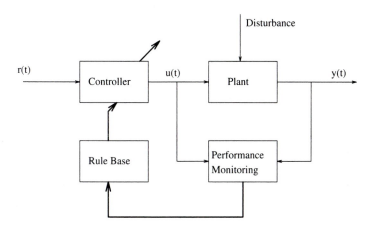

Figure 1.3: Rule based adaptive control system.

1.2.5 Artificial neural networks and fuzzy logic

The recently intensively studied artificial neural networks and fuzzy logic brings a new stage in the development of adaptive control, particularly for unknown nonlinear systems.

The idea of the artificial neural network is not new, the first can be traced to more than 30 years ago, [169]. However, neural network applications to control system design are new and have only been intensively studied over the past five years. Many types of artificial neural networks have been developed since 1980. Artificial Neural Networks (ANN) are connectionist algorithms [169] whose purpose is to approximate a nonlinear multivariate input/output mapping by utilising experimental data. For modelling and control applications, most ANNs studied have been feedforward networks comprised of multiple layers of transfer functions with connection weights which are adjusted by output error feedback to generate the desired input-output relationship. The network topology is primarily defined by the transfer functions utilised in the network, these may be either global (such as sigmoidials, gaussians, etc) or local (B-splines, radial basis functions, see chapter 3 and 4). ANNs require these issues to be addressed - the network architecture, the network performance criteria, and the training rules used to adjust the networks connection weights. It has been shown, after the training phase, that most artificial neural networks can be used to approximate nonlinear mappings to a specified accuracy.

If the nonlinear mapping considered is the relationship between the input and the output of a nonlinear unknown plant, an appropriate neural network can be used to model the plant. The obtained model is referred to as a neuro model for the plant. Since neural networks are well defined mathematically, the neuro model can also be regarded as a mathematical model for the plant. Using the neuro model, a controller can then be constructed by inverse model synthesis, model reference, label-feature generation or any appropriate nonlinear controller design method. The whole procedure, the training and the construction of the controller, can be either on-line or off-line [19, 169]. In the on-line framework, the neuro model is updated after each training data (i.e., the measured plant input and output pair) is received and the controller is directly evaluated using the updated neuro model. This is quite similar to the self-tuning adaptive control for linear systems, where the updating phase of neural model can be regarded as an on-line system identification. In off-line cases, the trained neural network is taken as the system model. The controller design simply uses the model and no adaptive updating is involved. Figure 1.4 shows the general structure of artificial neural network based adaptive control systems.

Artificial neural network based adaptive control has found increasing favour in many aspects of industrial process control. It belongs to the scope of advanced adaptive control and many outstanding problems still need to be resolved, such as stable closed loop control system design and the evaluation of control system performance. Many control schemes have been developed, but they still largely depend upon the particular structure of the plant, such as the control design for unknown affine nonlinear systems (see chapter 4). Therefore, unified approach to the improved closed loop control for nonlinear unknown systems will be an important topic in the future.

Fuzzy logic offers an alternative approach, when the plant to be controlled cannot be modelled by any form of numeric or algorithmic model or *a priori* data is in the form of production rules. In this case, a set of rules are used to evaluate control output. As is

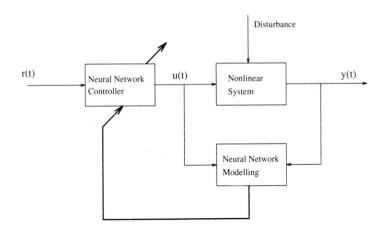

Figure 1.4: Neural networks based adaptive control system.

shown in Fig. 1.5, the whole system is normally composed of the plant to be controlled, a fuzzification unit (F) which fuzzifies the measured output of the plant, a controller and a defuzzification unit (DF) which generates a quantitative output from the controller. However, the input/output relationship of fuzzy logic controllers is entirely deterministic, it is only the data or knowledge that is represented by fuzzy variables.

The rule set can be divided into two subsets; the rules about the plant; and secondly, the rules about the controller. A fuzzy logic based control system is adaptive or self-organizing if the rules about the plant are updated on-line in response to a performance index evaluation. Since fuzzy sets are characterized by membership functions, the actual output to the plant is the weighted combination of some relevant outputs from the fuzzy sets. The degree of the contribution of each fuzzy set is determined by the value of its membership function. This is the major difference between the rule-based adaptive tuning where only one rule will be effective at a time. It has been shown recently [19] that a fuzzy logic based controller is input/output equivalent to an artificial neural network based controller, the significance of this is that all the mathematical analysis and training properties of a class of ANNs applies equally to the linguistic or symbolically based fuzzy logic process, providing an environment for fusing numeric and qualitative knowledge. Detailed discussion will therefore be given in chapters 3 and 4, where artificial neural network and fuzzy logic based modelling and control will be particularly addressed.

1.2.6   Application of adaptive control

Whilst many of the more esoferic theories of modern control theory such as H_∞ control, characteristic loci have found few practical applications that warrant their use, adaptive

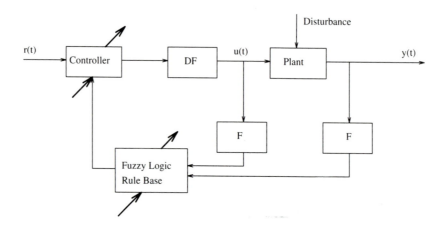

Figure 1.5: A typical fuzzy logic based control system.

control has been applied to many systems and has successfully solved many control problems that have been very difficult to solve using other methods. The following is, among numerous others, a list of some major application areas of adaptive control [118, 63]:

- Airspace control systems

- Electrical power control systems

- Chemical process control systems

- Autonomous guided vehicle control

- Industrial robots control

- Steel mill control

- Paper making process control

- Temperature and humidity control

- AC/DC motor speed control systems

- Medical treatment.

The increasing demand for high peformance, full operating envelope use will require the development of provable nonlinear adaptive controllers - this is partially addressed in this book.

1.3 The Context of Adaptive Control

The general class of dynamical processes considered in this book are finite dimensional and may be represented in the discrete time format:

$$x(k+1) = g(x(k), u(k), e_1(k), k) \qquad (1.3.1)$$
$$y(k) = h(x(k), e_2(k), k) \qquad (1.3.2)$$

where $x(k)$, $y(k)$, $u(k)$ are the system state, output and input respectively, $\{e_1(k), e_2(k)\}$ are noise or disturbance terms or represent mismodelling errors, also the nonlinear functions $g(.)$, $h(.)$ dependency on k indicates that the process may be time varying. Generally as only input/output data is available in adaptive control, the system model equations (1.3.1)-(1.3.2) are represented by the multivariable input-output model

$$y(k) = f(y(k-1), y(k-2), \cdots, y(k-n_y), u(k-d), u(k-d-1), \cdots, u(k-d-n_u)) + e(k) \qquad (1.3.3)$$

where $y \in R^m, u \in R^r, e \in R^m, n = (mn_y + rn_u + 1)$, and $f(.) : C(D) \in R^n \to R^m$, is an unknown vector valued nonlinear function defined on D, a compact subset of R^n. (n_u, n_y) represent the order of the dynamics of the representation (1.3.3), d is the time delay. By defining the information or measurement or observation vector as

$$\xi(k) = (y^T(k-1), \cdots, y^T(k-n_y), u^T(k-d), \cdots, u^T(k-d-n_u))^T \qquad (1.3.4)$$

then the system (1.3.3) can be represented simply as

$$y(k) = f(\xi(k)) + e(k) \qquad (1.3.5)$$

Hence as a modelling or identification problem the parameters (n_y, n_u, d) as well as $f(.)$ have to be found from the measurement data set $\{\xi(k)\}_{k=1}^N$ (see chapters 3-5). Given that the process model ((1.3.1)-(1.3.3)) has been estimated, a controller has to be synthetised so as to satisfy some performance criteria such as the cost

$$I(k) = \sum_{r=k}^N H(x(r+k), u(r), r) \qquad (1.3.6)$$

where $H(.)$ is a scalar cost function dictating the cost at time r, which depends on a time interval at k, the current control $u(r)$, and the state $x(r+k)$ consequents on that control. Minimisation of Eq. (1.3.6) subject to Eq. (1.3.1), recognising that $\{e_1, e_2\}$ are random variables, implies that $\{x, y, u, H\}$ are also random. Therefore the average cost $J = E(I(k))$ has to be evaluated and minimised in practice. The optimal value of J, J^*, is generally unavailable, as it is nonlinear and dependent upon complex boundary conditions. The class of problems for which it is mathematically tractable is when $H(.)$ is quadratic, $\{e_1, e_2\}$ are Gaussian distributed processes and $g(.)$, $h(.)$ are linear, resulting in the famous Linear Quadratic Gaussian (LQG) linear feedback controller [109, 161]. Almost all other conditions for $H(.)$, $g(.)$, $h(.)$ will lead to sub-optimal or nonanalytical solutions. In this book we shall consider the special case of Eq. (1.3.3) in the context of self-tuning regulators, when the input/output model is linear

$$\sum_{i=0}^{n_y} a_i y(k-i) = \sum_{j=0}^{n_u} b_j u(k-d-j) + \sum_{r=0}^{n} c_r e(k-r) + b \qquad (1.3.7)$$

where b is the system bias (usually assuemd zero). Any controller design based on the model (1.3.7) has the same degree of complexity and depends on the knowledge of the plant coefficients $\{a_i, b_j, c_r\}$ as well as the structural parameters $\{n_y, u_u, d, b\}$. If these are not available, they must be identified off-line or online via a simultaneous identification and control scheme. In the latter case real time estimation is necessary to generate the co-efficient $\{a_i, b_j, c_r\}$ in conjunction with a model order determination algorithm. This dual estimation process has to find a compromise between model complexity and model error, since the complexity of consequent model dictates controller synthesis and its complexity. Although the model (1.3.7) is linear in observations, it can become nonlinear when the coefficients $\{a_i, b_j, c_r\}$ are uncertain (as is usual in onlinear estimation). For example, a non-linearity is introduced by uncertainty in the model 'gain' and 'zeros' coefficients b_j, and in the 'pole' coefficients a_i, whereas uncertainty in the noise coefficients c_r, introduces a full nonlinear phenomena, complicating both estimation and control. In general adaptive control is associated with those problems for which $\{a_i, b_j, c_r\}$ are uncertain, leading to interaction between estimation and control, i.e., the certainty equivalence principle is not valid [12].

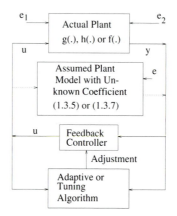

Figure 1.6: Adaptive controller structure.

For the special case when $g(.)$, $h(.)$ are assumed linear, but with unknown coefficients

the adaptive scheme of Fig. 1.6 represents the class of self-tuning and model reference controllers for which the generated control is based upon certainty equivalence and is suboptimal. In the case of self-tuning control [8, 9, 26], it is generally assumed that there is no uncertainty in the noise coefficients c_r, resulting in a linearised parameter identification problem. For example if $c_0 = a_0 = 1, b = 0, c_r = 0$, and

$$\theta = [-a_1, -a_2, \cdots, -a_{n_y}, b_0, b_1, \cdots, b_{n_u}]^T \tag{1.3.8}$$

represents the unknown parameter vector of model (1.3.7), then utilizing the information vector $\xi(k)$ in Eq. (1.3.4) gives the alternative representation

$$y(k) = \theta^T \xi(k) + e(k) \tag{1.3.9}$$

which is linear in the unknown parameter coefficients θ (assumed constant), which may be readily estimated recursively. Suppose that there are N observations of input/output data pairs $\{y(k), u(k)\}$, then (1.3.9) can be written in vector form as

$$\underline{y}_N = \underline{\xi}_N \theta + \underline{e}_N \tag{1.3.10}$$

where

$$
\begin{aligned}
\underline{y}_N &= (y^T(0), y^T(1), \cdots, y^T(N))^T \\
\underline{\xi}_N &= (\xi^T(0), \xi^T(1), \cdots, \xi^T(N))^T \\
\underline{e}_N &= (e^T(0), e^T(1), \cdots, e^T(N))^T
\end{aligned} \tag{1.3.11}
$$

If a quadratic cost function

$$J = N^{-1} \sum_{k=0}^{N} (y(k) - \xi^T(k)\theta)^T (y(k) - \xi^T(k)\theta) \tag{1.3.12}$$

is used, then the least mean squared (LMS) estimate of θ (see Chapter 3) is

$$\hat{\theta} = (\underline{\xi}_N^T \underline{\xi}_N)^{-1} \underline{\xi}_N^T \underline{y}_N \tag{1.3.13}$$

Additionally if $e(k)$ is a white noise sequence with variance σ^2, then $\hat{\theta}$ is a minimum variance estimate with covariance matrix

$$E((\theta - \hat{\theta})^T (\theta - \hat{\theta})) = \sigma^2 (\underline{\xi}_N^T \underline{\xi}_N)^{-1} \tag{1.3.14}$$

The parameter estimator (1.3.13) is in the form of a batch estimation where memory and computation cost increase with N. Clearly for online adaptive control this is highly undesireable and recursive one-step at a time estimators of the form new estimate - old estimate = learning rate × estimation error are frequently used (see chapters 3-7 for instantaneous learning laws), i. e.,

$$\Delta\hat{\theta}(k) = \mu(k)(y(k) - \xi^T(k)\hat{\theta}(k)) \tag{1.3.15}$$

Stability and convergence of this recursive algorithm depend on the learning rate or gain $\mu(k)$. In the case of the Kalman filter, $\mu(k)$ is a function of the estimation covariance matrix, and produces optimal estimation under certain conditions on model (1.3.7).

Clearly one of the fundamental issues of adaptive control is how to extend the theory, within a unified framework of existing theory, to nonlinear processes. Chapter 3,4 and 5 of this book address this issue by developing a neurofuzzy theory for modelling and control for classes of nonlinear dynamic processes. This is far from a complete theory, but does offer a continum to exsiting adaptive control theory. Similarly the classical self-tuning and model reference controller schemes for linear dynamic systems are extended in chapters 6-8, the operating point dependent scheme in chapter 5 for nonlinear processes can be readily extented to cover the cases of chapters 6-8.

This book, as its title suggests, is not intended as a first time reader for adaptive control, but rather as an advanced companion to references [12, 57] and [63].

1.4 Book Outline

The book contents are arranged as follows.

Chapter 2 contains a collection of assumed results and concepts from linear algebra, linear system, nonlinear systems and adaptive control, which are used throughout this book.

The modelling of nonlinear dynamical process is fundamental to nonlinear indirect adaptive control, the classical theory of polynomial (global and local) approximation theory offers viable solutions if some schemes can be found for adjusting the polynomial parameters. Feedforward supervised artificial neural networks offer this opportunity by providing arbitrary accuracy in functional approximation to $f(.)$ in Eq. (1.3.3). In Chapter 3, globally approximating and generalising networks such as the multilayer perceptron (MLP) and locally approximating and generalising networks such as the Radial Basis Function (RBF) and B-spline network are discussed. It is shown that local networks are appropriate to online learning with provable convergence conditions, and have the capacity to incorporate *a priori* knowledge. Issues of network architecture, functional approximation and learning laws are discussed in detail, whilst networks such as B-spline and RBF are appropriate to online learning, they suffer from the curse of dimensionality, therefore an introduction to the new field of parsimonions network construction algorithm is provided.

Fuzzy logic appears to have developed as a separate discipline from that of neural networks, the B-spline network provides powerful mechanism for interrelating the two fields into a single coherant field of neurofuzzy systems [19]. Chapter 4 provides a brief introduction to fuzzy logic and its associated operators and neurofuzzy adaptive modelling.

A novel approach is presented in chapter 5 for the modelling and control of a specific class of nonlinear systems whose linear parameters are unknown nonlinear functions of the measurable operating points. To identify each nonlinear function, a neural network is used with its input variables being the measurable operating points and its output being the estimated value of the parameter. Two different cases are considered in the modelling, the first being systems for which the neural networks can exactly model the nonlinear functions and the second case considers systems where the neural networks can only approximate the nonlinear functions to a known accuracy. The first type of systems is referred to as a matching system and the second type of systems is called a mismatching system. During the modelling phase, the weights for each neural network are trained in parallel and the normalized back-propagation (NBP) algorithm is used for matching systems and the modified recursive least square (RLS) algorithm is used for mismatching systems. It has been shown that this type of algorithm, together with the use of the technical lemma developed by Goodwin *et, al* [56, 57], will eventually lead to a stable d-step-ahead adaptive control scheme for matching systems and a pole assignment adaptive control strategy for mismatching systems.

An alternative approach, fuzzy logic based modelling, is also discussed in this chapter. In this case, a number of fuzzy logic units in chapter 4 are used to represent the nonlinear coefficients and the input to these fuzzy logic units are, again, the measurable operating points in linguistic form. The training algorithm for the fuzzy logic rule confidence matrix is discussed and its application to the control of a paper machine is included.

In Chapter 6 self-tuning sup regulators for discrete systems subjected to bounded increment external inputs are presented. Two kinds of input spaces are introduced to model the external environment of discrete systems, which can cover transient and persistent external inputs. The sup controllers minimize the supremum of the absolute value of the generalized output for all time and for all external inputs in one of the input spaces. The analysis, which is restricted to minimum phase discrete systems, yields the sup regulator for systems with disturbance but without reference input, establishes some relations between an input space, a sup controller and output performance and considers the stability of the closed-loop system. The sup regulators for non-minimum phase discrete systems are also studied. It is shown that they can be designed by numerical methods, e.g. minimax optimization. A self-tuning sup regulator algorithm based on the MLS algorithm is presented and analyzed. The global convergence analysis shows that the closed-loop system is stable and the properties of the self-tuning sup regulator asymptotically converge to those of the sup regulator. The results of a simulation example demonstrate the operation of the self-tuning sup regulators.

In Chapter 7 a self-tuning mean regulator for stochastical discrete systems is studied in parallel to the self-tuning sup regulator described in chapter 6. Another stochastic input space, rather than the one for white noises, is introduced to model the external environment of stochastic control systems. The mean regulator minimizes the supremum of the mean of the absolute value of the generalized output with respect to all external inputs in the input space. Again, the analysis yields the design procedure of the mean regulator and establishes some relations between the input space, the mean controller and the output performance of the system. Again, like chapter 6, stability of closed-loop systems is discussed. A self-tuning mean regulator algorithm based on both the MLS algorithm and the mean regulator is presented and analyzed. The global convergence analysis shows that the closed-loop system is stable and the properties of the self-tuning mean regulator asymptotically converge to those of the mean regulator. These results are illustrated by some simulation examples.

Chapter 8 presents a model reference adaptive predictive control (MRAPC) for time-varying systems with a variable time delay. After analyzing the structure of the well known Smith predictor control, two model reference adaptive predictive control strategies are discussed on the basis of a parametric optimization technique and Lyapunov stability theory. The MRAPC strategy consists of a reference model, an adaptive predictor, an adaptive controller and an adaptation mechanism. The adaptive controller is designed to ensure that the plant output tracks that of the reference model asympototically, whilst the adaptive predictor is used to predict the effect of the current control action on the actual process output. It has been shown that the MRAPC is much more effective than the Smith predictor control in reducing the influence of time-varying parameters and variable time delay on the system. Some simulation examples are included to illustrate the use of the MRAPC strategies.

Chapter 9 describes a recently developed rule based identification algorithm and rule based controller tuning. This forms a rule based adaptive control and can be regarded as another aspect in advanced adaptive control. After the discussion on rule based identification, it is directly used to identify the model between the reference step input and the integration of the tracking error. Two stages of tuning are introduced, in the first stage, the correct structure of the controller is chosen by monitoring the integral of the tracking error. This enables the realization of perfect steady state tracking. In the second stage, the estimated model is used to tune the controller gains in order to provide an improved dynamic performance for the closed-loop response. Applications of the algorithm to a number of real processes are included.

Chapter 10 presents a general adaptive control scheme for unknown time-invariant singular systems of the form $Ex(t + 1) = Ax(t) + Bu(t); y(t) = Cx(t)$. Due to the non-causality of such systems, the identification of unknown parameters is reconsidered and a new residual signal is constructed and used in a recursive estimation algorithm. Based on the parameter estimation, a general design procedure is obtained which includes the following two steps: i) preliminary output feedback gain calculation in order to make the original system causal; and ii) adaptive control design for the resulting causal system. It is shown that any adaptive control algorithm can be combined with this scheme to obtain a globally stable closed-loop system. A simulation using a third order singular system demonstrates the suitability of the design procedure, which is followed by an intensive discussion on the application of the algorithm to a gas turbine control system.

Each chapter is relatively independent, therefore readers may design their own reading patterns.

Chapter 2

Preliminaries

This chapter outlines some fundamental results about matrices, dynamic system representation, least square estimation and the stability analysis for adaptive control systems, that are necessary mathematical prerequested for this book.

2.1 Matrices

A matrix A of dimension $m \times n$ is defined as

$$A = [a_{ij}] = \begin{bmatrix} a_{11} & a_{12} & \cdots & a_{1n} \\ a_{21} & a_{22} & \cdots & a_{2n} \\ \vdots & \vdots & \ddots & \vdots \\ a_{m1} & a_{m2} & \cdots & a_{mn} \end{bmatrix} \quad (2.1.1)$$

where a_{ij} is the ijth element of A, m is the number of rows and n is the number of columns. If $m = n$, the matrix is referred to as a square matrix. Let $B = [b_{ij}]$ be a $p \times q$ matrix, then the addition of A and B, only when $p = m$ and $q = n$, is defined as

$$C = A + B = [a_{ij} + b_{ij}] \quad (2.1.2)$$

Example 2.1 Let

$$A = \begin{bmatrix} 1 & 2 & -1 \\ 0 & -4 & 6 \end{bmatrix}; B = \begin{bmatrix} 2 & -1 & 0 \\ -1 & 7 & 3 \end{bmatrix}$$

then

$$C = A + B = \begin{bmatrix} 3 & 1 & -1 \\ -1 & 3 & 9 \end{bmatrix}$$

Only in the case that $p = n$, the multiplication of A and B can be defined and the resulting matrix C, of dimension $m \times q$, is given by

$$C = AB = [c_{ij}] \quad (2.1.3)$$

$$c_{ij} = \sum_{k=1}^{n} a_{ik} b_{kj} \quad (2.1.4)$$

Using the above definitions, the following properties for fundamental matrices calculation can be directly obtained:

$$A + B = B + A \tag{2.1.5}$$
$$ABC = (AB)C = A(BC) \tag{2.1.6}$$
$$A(B + C) = AB + AC \tag{2.1.7}$$
$$(B + C)A = BA + CA \tag{2.1.8}$$

In general $AB \neq BA$. The transpose of $A = [a_{ij}]$ is defined as

$$A^T = \begin{bmatrix} a_{11} & a_{21} & \cdots & a_{m1} \\ a_{12} & a_{22} & \cdots & a_{m2} \\ \vdots & \vdots & \ddots & \vdots \\ a_{1n} & a_{2n} & \cdots & a_{mn} \end{bmatrix} \tag{2.1.9}$$

Matrix A of dimension nx1 is referred to as an n-dimensional vector and is denoted by $A \in R^n$, its transpose is defined as a n-dimensional row vector. A square matrix $A \in R^{n \times n}$ can have the determinant [13] which is denoted by $det(A)$, A is therefore called a nonsingular matrix if $det(A) \neq 0$. For the nonsingular matrix A, its inverse is denoted by A^{-1} and is the unique solution of

$$AA^{-1} = A^{-1}A = I_n \tag{2.1.10}$$

where I_n is the unit matrix with its diagonal elements equal to 1 and off-diagonal elements equal to 0.

2.2 Eigenvalues, Eigenvectors and Eigenrows

The eigenvalues of a square matrix, A, are defined as the roots of the following polynomial

$$f(\lambda) = det(\lambda I - A) \tag{2.2.1}$$

Denote $\lambda_1, \lambda_2, ..., \lambda_n$ as the eigenvalues of A, the corresponding eigenvector $v_i \in R^n$ for the ith eigenvalue λ_i is the nonzero solution of

$$Av_i = \lambda_i v_i \tag{2.2.2}$$

and the corresponding eigenrows $w_j \in R^{1 \times n}$ for the jth eigenvalue λ_j is the nonzero solution of

$$w_j A = \lambda_i w_j \tag{2.2.3}$$

Example 2.2 Let

$$A = \begin{bmatrix} 0 & 1 \\ -2 & -3 \end{bmatrix}$$

then
$$f(\lambda) = det(\lambda I - A) = \lambda^2 + 3\lambda + 2$$

and the two eigenvalues are
$$\lambda_1 = -1; \qquad \lambda_2 = -2$$

The eigenvector corresponding to λ_1 is

$$v_1 = \begin{bmatrix} 1 \\ -1 \end{bmatrix}$$

and the eigenrow for λ_1 is
$$w_1 = \begin{bmatrix} 2 & 1 \end{bmatrix}$$

2.3   Time-varying Matrices

A matrix or vector is time-varying if one or more than one of its elements is a function of time t. The derivative of a matrix A with respect to t is then defined as

$$\frac{dA}{dt} = [\frac{da_{ij}(t)}{dt}] = \begin{bmatrix} \frac{da_{11}}{dt} & \frac{da_{12}}{dt} & \cdots & \frac{da_{1n}}{dt} \\ \frac{da_{21}}{dt} & \frac{da_{22}}{dt} & \cdots & \frac{da_{2n}}{dt} \\ \vdots & \vdots & \ddots & \vdots \\ \frac{da_{m1}}{dt} & \frac{da_{m2}}{dt} & \cdots & \frac{da_{mn}}{dt} \end{bmatrix} \tag{2.3.1}$$

and the following properties hold.

$$\frac{d(A(t)B(t))}{dt} = (\frac{dA(t)}{dt})B(t) + A(t)(\frac{dB(t)}{dt}) \tag{2.3.2}$$

$$\frac{dA^{-1}(t)}{dt} = -A^{-1}(t)(\frac{dA(t)}{dt})A^{-1}(t) \tag{2.3.3}$$

Example 2.3 Assume that
$$A = \begin{bmatrix} sin(t^2) & 1 \\ e^{-t} & t \end{bmatrix}$$

then
$$\frac{dA(t)}{dt} = \begin{bmatrix} 2tcos(t^2) & 0 \\ -e^{-t} & 1 \end{bmatrix}$$

2.4   Norms and Inner Products

Assume that $x \in R^n$ be a vector, then its norm $\|x\|$ is defined as

$$\|x\| = [\sum_{i=1}^{n} x_i^2]^{1/2} \tag{2.4.1}$$

where x_i is the ith component of x.

Example 2.5 Consider

$$x = \begin{bmatrix} -4 \\ 0 \\ 3 \end{bmatrix}$$

then we have

$$\|x\| = [(-4)^2 + 0^2 + 3^2]^{1/2} = 5$$

The norm of a vector can be interpreted as a measure of its size or length and the following properties must therefore be satisfied.

i) $\|x\| > 0$ for any x, $\|x\| = 0$ if and only if $x = 0$;

ii) For any scalar a, $\|ax\| = |a| \, \|x\|$;

iii) $\|x + y\| \leq \|x\| + \|y\|$ for all x and y.

Other widely used norms are given below

$$\|x\|_\infty = \max_i |x_i| \qquad (2.4.2)$$

$$\|x\|_1 = \sum_{i=1}^{n} |x_i| \qquad (2.4.3)$$

where the first one is referred as H^∞ norm whilst the second one is called l_1 norm. They are all very useful norms for control system design.

Since a matrix $A \in R^{m \times n}$ can be regarded as an mn dimensional vector, its norms can be similarly defined as those for vectors. A commonly used matrix norm is called an *induced* norm which is defined as

$$\|A\| = \max_{\|y\|=1} \|Ay\| \qquad (2.4.4)$$

where $y \in R^n$ is a column vector. For matrix norms, the three properties mentioned earlier should also apply. Moreover, we have

$$\|Ay\| \leq \|A\| \, \|y\| \qquad (2.4.5)$$
$$\|AB\| \leq \|A\| \, \|B\| \qquad (2.4.6)$$

where y is a vector and B is another matrix.

Let $x, y \in R^n$ be any two vectors, its inner product is a scalar and is defined as

$$< x, y >= x^T y = \sum_{i=1}^{n} x_i y_i \qquad (2.4.7)$$

where x_i and y_i are the ith component of x and y, respectively.

Example 2.6 Assume

$$x = \begin{bmatrix} 1 \\ 0 \\ 3 \end{bmatrix} ; y = \begin{bmatrix} 0 \\ 10 \\ 2 \end{bmatrix}$$

then we have

$$< x, y >= 1 \times 0 + 0 \times 10 + 3 \times 2 = 6$$

It can be seen that

$$< x, x >= \|x\|^2 \tag{2.4.8}$$

For the inner product, the following Schwartz inequality holds.

$$|< x, y >| \le \|x\| \, \|y\| \tag{2.4.9}$$

2.5 Dynamic System Representations

2.5.1 Continuous time systems

Using matrices and vectors, dynamic systems can be expressed in simple forms. For a nth order linear system, the following state space representation

$$\dot{x}(t) = Ax(t) + Bu(t) + Ew(t) \tag{2.5.1}$$
$$y(t) = Cx(t) + Du(t) + Fw(t) \tag{2.5.2}$$

are often used, where $x(t) \in R^n$ is the state vector, $u(t) \in R^m$ is the measured input vector, $y(t) \in R^p$ is the measured output and $w(t) \in R^l$ is the unmeasurable external disturbance vector which can be either a deterministic or a stochastic process. A, B, C, D, E and F are parameter matrices of dimensions $n \times n$, $n \times m$, $p \times n$, $p \times m$, $n \times l$ and $p \times l$, respectively.

The system is linear time-invariant (LTI) if all the parameter matrices have constant elements, otherwise, it is referred to as the time-varying linear system. For nonlinear dynamic systems, the state space representation becomes

$$\dot{x}(t) = f(x(t), u(t), w(t), t) \tag{2.5.3}$$
$$y(t) = g(x(t), u(t), w(t), t) \tag{2.5.4}$$

where $f(...)$ and $g(...)$ are nonlinear unknown or known functions, and $x(t), u(t), w(t)$ and $y(t)$ have the same meanings as those for linear systems. A class of nonlinear systems, which are often referred to as affine systems in literature, are specially expressed as

$$\dot{x}(t) = f(x(t)) + g(x(t))u(t) \tag{2.5.5}$$
$$y(t) = h(x(t)) \tag{2.5.6}$$

In many practical cases the state vector of dynamic systems is difficult to obtained, input-output expressions are widely adopted as the starting point for control system design and implementation. For linear time-invariant systems the following expression, often called transfer function or matrix description, is used.

$$
\begin{align}
Y(s) &= G_1(s)U(s) + G_2(s)\Omega(s) \tag{2.5.7}\\
G_1(s) &= C(sI - A)^{-1}B + D \tag{2.5.8}\\
G_2(s) &= C(sI - A)^{-1}E + F \tag{2.5.9}
\end{align}
$$

where $Y(s)$, $U(s)$ and $\Omega(s)$ are the Laplace transforms of time domain functions $y(t)$, $u(t)$ and $w(t)$, respectively.

2.5.2 Discrete time systems

Since most practical adaptive control systems are implemented via digital computers, the discrete-time representation for dynamic systems is critical. There are two different ways by which a continuous-time system is discretised, either use of q^{-1} operator or via the δ operator [110]. In this book only the q^{-1} operator will be used. Denote T_0 as the sampling time, then the sampled values of system input and output are $u(kT_0)$ and $y(kT_0)$ with $k = 0, 1, 2, \cdots$. To simplify the notation in the consequent chapters, T_0 will be omitted. The q^{-1} operator is defined as

$$
q^{-1}y(k) = y(k-1) \equiv y(kT_0 - T_0) \tag{2.5.10}
$$

As a result, the following state space difference equation

$$
\begin{align}
x(k+1) &= Ax(k) + Bu(k) + Ew(k) \tag{2.5.11}\\
y(k) &= Cx(k) + Du(k) + Fw(k) \tag{2.5.12}
\end{align}
$$

or the input-output expression

$$
\begin{align}
A(q^{-1})y(k) &= q^{-d}B(q^{-1})u(k) + C(q^{-1})w(k) \tag{2.5.13}\\
A(q^{-1}) &= 1 + a_1 q^{-1} + a_2 q^{-2} + \ldots + a_n q^{-n} \tag{2.5.14}\\
B(q^{-1}) &= b_0 + b_1 q^{-1} + \ldots + b_m q^{-m} \tag{2.5.15}\\
C(q^{-1}) &= c_0 + b_1 q^{-1} + \ldots + c_l q^{-l} \tag{2.5.16}
\end{align}
$$

is used to describe discrete-time linear time-invariant systems (DTLTIS), where d represents the time delay. Eq. (2.5.13) is refereed to as Auto-Regression and Moving Average with eXogent model (ARMAX) for DTLTIS, which has been widely utilized in the modelling and control literature.

Example 2.7 The following system

$$
y(k) = -1.5y(k-1) + 0.56y(k-2) + 0.06u(k-4)
$$

can be expressed in unit shift operator q^{-1} as follows

$$(1 + 1.5q^{-1} - 0.56q^{-2})y(k) = 0.06q^{-4}u(k)$$

It can be seen that

$$A(q^{-1}) = 1 + 1.5q^{-1} - 0.56q^{-2}$$
$$B(q^{-1}) = 0.06$$

For nonlinear systems, the discrete state space expression becomes

$$
\begin{align}
x(k+1) &= f(x(k), u(k), \omega(k), k) \tag{2.5.17}\\
y(k) &= g(x(k), u(k), \omega(k), k) \tag{2.5.18}
\end{align}
$$

and the input-output form is

$$
\begin{align}
y(k) = {} & f(y(k-1), y(k-2), ..., y(k-n), u(k-d), u(k-d-1), ..., \\
& u(k-d-m), \omega(k), \omega(k-1), ..., \omega(k-l)) \tag{2.5.19}
\end{align}
$$

where $f(...)$ and $g(...)$ are nonlinear function which may be either known or unknown, n, m and l are integers representing the structure orders of the system. Eq. (2.5.19) is often referred to as a Nonlinear Auto-Regression and Moving Average with eXogent model (NARMAX) for nonlinear systems. Many neural network based adaptive control schemes use NARMAX model to express the system, where artificial neural networks are employed to estimate the unknown nonlinear function $f(\cdots)$.

2.6   Least Square Estimation

The least square estimation, especially its recursive form, has widely been used in self-tuning adaptive control over the past few decades. In this section, we will briefly introduce the algorithm and more details are available in many sources [57].

Assume that there is no disturbance in Eqs. (2.5.13)-(2.5.16) (i.e., $\omega(t) = 0$) and denote

$$
\begin{align}
\theta &= (-a_1, -a_2, ..., -a_n, b_0, b_1, ..., b_m)^T \in R^{n+m+1} \tag{2.6.1}\\
\phi(k-d) &= (y(k-1), y(k-2), ..., y(k-n), u(k-d), \\
& \quad u(k-d-1), ..., u(k-d-m))^T \in R^{n+m+1} \tag{2.6.2}
\end{align}
$$

as the parameter (unknown) and observation vectors, then it can be seen that

$$y(k) = \theta^T \phi(k-d) \tag{2.6.3}$$

Let

$$\hat{\theta}(k) = (-\hat{a}_1(k), -\hat{a}_2(k), ..., -\hat{a}_n(k), \hat{b}_0(k), \hat{b}_1(k), ..., \hat{b}_m(k))^T \in R^{n+m+1} \tag{2.6.4}$$

be the estimate of θ at $t = kT_0$, then the known recursive least square algorithm is given by

$$\Delta\hat{\theta}(k) = \hat{\theta}(k) - \hat{\theta}(k-1) = \frac{P(k-2)\phi(k-d)\epsilon(k)}{1 + \phi^T(k-d)P(k-2)\phi(k-d)} \qquad (2.6.5)$$

$$\epsilon(k) = y(k) - \hat{y}(k) \qquad (2.6.6)$$

$$\hat{y}(k) = \hat{\theta}^T(k-1)\phi(k-d) \qquad (2.6.7)$$

$$P^{-1}(k-1) = P^{-1}(k-2) + \phi(k-d)\phi^T(k-d) \qquad (2.6.8)$$

with $\hat{\theta}(0) = 0$ and $P(0) = P^T(0) > 0$ being initial values.

The convergence of the above algorithm is established by the following theorem.

Theorem 2.1 *When the algorithm (2.6.5)-(2.6.8) is applied to the input and output data $\{u(k), y(k)\}$ of system (2.6.3), then*

$$\lim_{k\to\infty} \frac{\epsilon^2(k)}{1 + \phi^T(k-d)P(k-2)\phi(k-d)} = 0 \qquad (2.6.9)$$

$$\lim_{k\to\infty} \left\| \hat{\theta}(k) - \hat{\theta}(k-k_0) \right\| = 0 \qquad (2.6.10)$$

where k_0 is a positive integer.

The proof of the above theorem can be found in [57] and is therefore omitted here.

2.7 Technical Lemma

In chapter 1 we mentioned that various Lyapunov stability criteria were widely applied to the design of adaptive control strategies in order to stabilize the resulting closed loop system; for unknown discrete-time systems the technical lemma, of Goodwin *et, al* [56], can also be used to construct stable adaptive control systems. In this section, this lemma and its use will be particularly addressed.

Lemma 2.1 *Suppose that the following conditions are satisfied for some given sequences $\epsilon(k) \in R^1$, $\phi(k) \in R^n$, σ, $\sigma_1(k) \in R^1$ and $\sigma_2(k) \in R^1$:*
 1)

$$\lim_{k\to\infty} \frac{(\epsilon(k) - \sigma)^2}{\sigma_1(k) + \sigma_2(k)\phi^T(k)\phi(k)} = 0 \qquad (2.7.1)$$

 2)

$$0 < \sigma_1(k) < \infty \qquad (2.7.2)$$

$$0 < \sigma_2(k) < \infty \qquad (2.7.3)$$

3)

$$\|\phi(k)\| \leq c_1 + c_2 \max_{0 \leq \tau \leq k} |\epsilon(\tau)| \tag{2.7.4}$$

where c_1 and c_2 are positive finite numbers. Then

$$\lim_{k \to \infty} \epsilon(k) = \sigma \tag{2.7.5}$$

$$\|\phi(k)\| < \infty \tag{2.7.6}$$

The proof of this lemma (when $\sigma = 0$) can be found in [56, 57] and is, again, omitted here. Of these three conditions, it can be seen that condition 1) is similar to Eq (2.6.9) in theorem 2.1. The only difference here is that a scalar sequence, $\sigma_2(k)$, is used instead of the non-negative matrix $P(k-2)$. Therefore, recursive least square parameter estimation can be normally used to guarantee conditions 1)-2) when $\sigma = 0$. However, condition 3) is unusual and can not be naturally obtained from recursive parameter estimation. Rather, it should arise from the construction of the self-tuning adaptive controller. Since all the components in vector $\phi(k)$ are composed by the system input and output sequences, $u(k)$ and $y(k)$, the technical lemma described above can be used to construct the controller.

To summarize, the following two major procedures are obtained for the design of stable self-tuning adaptive control systems.

i) Construct a recursive parameter estimation algorithm so that the conditions 1) and 2) of the technical lemma 2.1 are satisfied;

ii) Establish a self-tuning adaptive controller in order to guarantee condition 3).

These two procedures have been frequently used in the convergence analysis of various adaptive control schemes in the book by Goodwin and Sin [57]. They will also be applied in chapter 5 and 6 in order to build up a neural network based controller for a class of nonlinear unknown systems and sup controllers for unknown linear systems.

2.8 A Design Example

In this section, the design of a stable one-step-ahead adaptive control system will be used in order to illustrate the two procedures described in section 2.7. Similar to section 2.6, we only consider a noise-free ARMA model, and in specific, we assume that the system has a unit time delay (i.e. $d = 1$) and that $m = 0$. The system can therefore be given by

$$y(k) = -a_1 y(k-1) - a_2 y(k-2) - \cdots - a_n y(k-n) + b_0 u(k-1) \tag{2.8.1}$$

where a_i and b_0 are unknown constant parameters. Using the same notations (2.6.1)-(2.6.2), the system can be further expressed as

$$y(k) = \theta^T \phi(k-1) \tag{2.8.2}$$

2.8.1 Parameter estimation

The recursive least square algorithm is used to identify vector θ. Using theorem 2.1, we have

$$\lim_{k \to \infty} \frac{\epsilon^2(k)}{1 + \phi^T(k-1)P(k-2)\phi(k-1)} = 0 \tag{2.8.3}$$

where

$$\epsilon(k) = y(k) - \theta^T(k-1)\phi(k-1) \tag{2.8.4}$$

Since

$$P(k-2) \le P(k-3) \le \cdots \le P(0) \le c_0 I_{n+1} \tag{2.8.5}$$

with I_{n+1} being the unit matrix and c_0 the maximum eigenvalue of matrix $P(0)$, it can be shown that

$$\lim_{k \to \infty} \frac{\epsilon^2(k)}{1 + c_0 \phi^T(k-1)\phi(k-1)} = 0 \tag{2.8.6}$$

This is actually a specific form of the conditions 1)-2) of the technical lemma, with $\sigma_1(k) = 1$ and $\sigma_2(k) = c_0$.

2.8.2 Control algorithm and stability

Let $y_r(k)$ be a bounded demand signal and consider the one-step-ahead controller which brings the output of the system to $y_r(k)$ in one step. Using the identification algorithm and the certainty equivalence principle, the control signal $u(k)$ is recursively calculated as follows

$$u(k) = \frac{1}{\hat{b}_0(k-1)}(y_r(k) + \hat{a}_1(k-1)y(k-1) + \cdots + \hat{a}_n(k-1)y(k-n)) \tag{2.8.7}$$

where it is also assumed that $\hat{b}_0(k-1) \ge \delta_1 > 0$ for any $k > 1$. Clearly, this expression can be re-written to give

$$y_r(k) = \theta^T(k-1)\phi(k-1) \tag{2.8.8}$$

As a result, Eq. (2.8.2) can be modified to the following form

$$y(k) = y_r(k) + \epsilon(k) \tag{2.8.9}$$

Since $y_r(k)$ is bounded, there exits $0 \le \delta_2 < \infty$ and $0 \le \delta_3 < \infty$ such that

$$|y(k)| \le \delta_2 + \delta_3 \max_{0 \le \tau \le k} |\epsilon(\tau)| \tag{2.8.10}$$

From Eq. (2.6.10) in theorem 2.1, it can also be concluded that $\hat{\theta}(k)$ is bounded. Therefore, the following inequality

$$|u(k)| \le \delta_4 + \delta_5 \max_{0 \le \tau \le k} |\epsilon(k)| \tag{2.8.11}$$

can be directly obtained by combining Eq. (2.8.7) and inequality (2.8.10), where δ_4 and δ_5 are, again, positive finite numbers. From the definition of vector $\phi(k-1)$ and inequalities (2.8.10) and (2.8.11), it can be shown that

$$\|\phi(k-1)\| \le c_1 + c_2 \max_{0 \le \tau \le k} |\epsilon(k)| \tag{2.8.12}$$

Using the technical lemma, it can be seen that

$$\lim_{k \to \infty} \epsilon(k) = 0 \tag{2.8.13}$$

and from Eqs. (2.8.10)-(2.8.11).

$$|u(k)| < \infty \qquad |y(k)| < \infty \tag{2.8.14}$$

Therefore, the closed loop system is stable and all the variables are uniformly bounded.

Chapter 3

Artificial Neural Networks: Aspects of Modelling and Learning

3.1 Introduction

Intelligent modelling and control research has enjoyed considerable attention in recent years; intelligent controllers are enhanced adaptive or self-organising controllers that adapt automatically to changes in the plant or its associated environment by observational or experiential data alone, without recourse to physical modelling. Intelligent Control (IC) is formulated from concepts and techniques in neurophysiology, operational research, control theory and artificial intelligence. Inherent to any control theory is the formulation or identification of appropriate process or plant models, in intelligent control, rather than utilise inappropriate process linear-physio-mathematical models, models are derived from plant input-output mappings from experimental evidence via an associative memory which may incorporate a priori process knowledge. The strength and weakness of this approach is that it does not depend on the restrictive but enabling assumption of a linear time invariant plant. Intelligent controllers determine their structure and consequent behaviour in response to their interaction with the environment and their current and past performance. Two apparently disparate methodologies have seen rapid growth in recent years - Artificial Neural Networks (ANN) and Fuzzy Logic Networks (FLN) [64, 65], both approaches make few restrictions on the type of input-output mapping that can be learnt. However despite the rapidly growing numbers of applications, the behaviour, properties and interrelationships of these algorithms is poorly understood. This is particularly true for FLN, where the application domain of range includes cameras, domestic products such as vacuum cleaners, washing machines, rice cookers, transportation, manufacturing etc. It is now measured in $ billion pa; but the decision surfaces are poorly understood, and few convergence and stability conditions exist - critical for any safety critical process. Chapter 4 considers the interrelationship between ANN and FLN, and in depth analytical study can be found in [19].

It would appear that IC implies that the system incorporates some form of neural or fuzzy algorithm [166]. However, as these algorithms become mathematically well understood and integrated into conventional control theory, it can be argued that they are no longer intelligent. It can be argued that its not the technique used that is intelligent but rather the application which defines an intelligent approach, and some leading AI researchers take this view in that it is application domains such as speech, vision, handwriting, adaptive planning etc., which requires an intelligent approach. Intelligent algorithms should possess the ability to:

- **sense** the surrounding events and world,

- **influence** their interaction with the environment, and

- **model** their cause and effect relationships,

but only because the applications require such an approach. The techniques used to implement the behaviour should also be able to:

- **plan** and reason about these actions,

- **learn** from their interaction with the environment,

- **generalise** information to similar situations, and

- **abstract** common concepts autonomously.

Some of the ANN and FLN considered in this chapter and chapter 4 have the ability to:

- **learn** autonomously,

- **incorporate** any prior knowledge about the system,

- **extract** knowledge about the process from the trained network,

- **abstract** relevant concepts that enable the network to generalise appropriately,

- **provable** behavioural characteristics such as arbitrary approximation capability, and learning convergence and stability.

These attributes do not satisfy all the requirements for intelligent behaviour. Indeed, the important network provability conditions, essential for online learning, diminish its perceived intelligence. Originally the opaque structure of an ANN was not thought to be a problem, although in practice many systems must go through an extensive training/verification/editing cycle in order to overcome some of the problems associated with incomplete training data. This is not possible unless the information learnt by the network is stored in a transparent manner. Networks in which a priori knowledge can be incorporated are inherently well conditioned, learning input/output relationships quickly, which is an important requirement for *online* modelling. Generally, the problems of learning online are greater than designing a network offline as many of the data gathering, validation and verification procedures (both for the network and the data) must be performed autonomously. Training an ANN offline usually involves determining which inputs are significant, which interactions should be modelled, and producing an appropriate internal representation automatically. Online and offline training are treated separately here; for online training, simple learning laws should be used to adapt appropriately structured networks and any offline training procedures should be used to develop this internal representation; abstracting concepts which are appropriate for both the network's approximation ability and its online learning laws.

The interest in IC lies in the need to control increasingly complex dynamical processes, whilst satisfying more stringent performance criteria, yet at the same time accommodating process temporal and spatial variations subject to disturbances and faults. Inherently, intelligent controllers are nonlinear and would appear to offer little to the control of linear processes for which there exist a wealth of theory. However, few processes are globally linear, frequently there are significant practical limits on control signals, imposition of static and dynamic constraints for process safety, and IC can offer better than optimal control for some processes [6]!

There are three important elements of any intelligent modelling algorithm or network:

i) the networks architecture or topology and associated interconnections,

ii) the networks functional approximation and representation ability, and

iii) the learning laws by which the network adapts itself to external influences or experiences.

Each of these issues and their interrelationships will be explored in this chapter predicated by their relevance to intelligent modelling and control. Generally the modelling capabilities of a network can be analysed separately from a training rule, thereby a variety of different adaptive learning rules (both supervised and unsupervised) and associated performance characteristics can be applied to the network's structure. Additionally as with any identification scheme, persistent and sufficient signal excitation is required over the whole input space domain, and the modelling error and measurement noise must be sufficiently well filtered to avoid over parameterisation (*i.e.* over parameterised models fit the noise, correctly parameterised models automatically filter noise).

IC can be categorised as static systems, in which there are fixed rule bases akin to a look up table (e.g. a static fuzzy logic controller, or an adaptive system in which fuzzy rules or associated basis functions are altered according to existent conditions. Similar to classical adaptive control, IC can be categorised as being of the direct or indirect type. In direct IC schemes [65] the control laws are synthesised without reference to any intermediate explicit parameterised process model, but rather is determined directly from observed errors; in practice a Jacobian of the plant is usually required to identify input/output causality.

Whereas in indirect IC schemes [65] an explicit appropriate process model is formed utilising the identification methods of this chapter, then various methodologies including model inversion are used to provide control. Several IC architectures are available, but are outside the scope of this book (see [19]).

Generally, neural networks are model free (or non-parametric), meaning that there are no a priori assumptions made about the form of the unknown input/output function that must be identified. In practice there is considerable a priori knowledge such as functional smoothness, common sense knowledge about the process, estimates of input/output time delays and the degree of correlation and causality between variables etc. The in-position of this a priori knowledge and constraints not only improves learning but also converts ill-posed approximation problems into well posed ones by means of regularisation. Neural networks can be viewed and interpreted from several viewpoints, and in this book they are developed in the context of classical numerical approximation and signal processing theory, since they share many common mathematical properties. In assessing various neural networks, their modelling properties need to be considered: can a network approximate a continuous nonlinear function with arbitrary accuracy, how does the network generalise (interpolate and extrapolate) between data, what class of functions can a fixed structure represent exactly, what are the resources required to implement the network? A similar set of questions can be posed about the ability of a neural network to *learn* the data: what are the best learning laws to ensure guaranteed parametric convergence and optimal performance, and how does the network's structure influence parametric convergence and noise rejection etc.? Some standard concepts, such as network *conditioning* (see section 3.7.3), can be used to provide partial answers to these questions, although much work still needs to be performed in these areas. In this chapter we shall assess the ability, and suitability, of neural networks to identify unknown dynamical processes based only on observed input/output data. It is beyond the scope of this book to treat neural networks in depth (see [19, 67, 80]) although it does compare and contrast different network topologies, the approximation capabilities, their computational and topological issues associated with the dimension of the "sensor" input space as well as the rapidly developing field of network construction algorithms and learning rules.

3.2   Network Approximation

Given an unknown, but observed, dynamical process with L input/output pairs of data $\{u(k), y(k)\}$, the general modelling problem is to find the functional dependence of y on u so as to:

  i) provide a description that enumerates the dependence of y on u,

  ii) allows inferences to be drawn on the relative contribution (influences/correlations) of the independent variables u on y, and

iii) predict values of y given sets of values of u.

These requirements are the same for most modelling techniques whether they are neural, fuzzy or linear time-invariant algorithms. The data should also contain enough information so that a computational algorithm can abstract an appropriate model, and this is a basic requirement for any learning system.

Consider the class of input-output nonlinear systems represented by

$$y(k) = f(y(k-1), \ldots, y(k-n_y), u(k-d), \ldots, u(k-d-n_u)) + e(k) \quad (3.2.1)$$

where

$$\begin{align}
y(k) &= (y_1(k), \ldots, y_m(k))^T & (3.2.2) \\
u(k) &= (u_1(k), \ldots, u_r(k))^T & (3.2.3) \\
e(k) &= (e_1(k), \ldots, e_m(k))^T & (3.2.4)
\end{align}$$

are the system output, input, and noise vectors respectively and $f() : C(D) \subset R^n \to R^m$ is a *unknown* vector valued nonlinear function defined on D which is a compact subset of R^n, $C(D)$ denotes a set of continuous real valued functions on D and $n = (mn_y + rn_u)$. From the definition of input information vector

$$\xi(k) = (y^T(k-1), \ldots, y^T(k-n_y), u^T(k-d), \ldots, u^T(k-d-n_u))^T \quad (3.2.5)$$

then system (3.2.1) can be represented simply as

$$y(k) = f(\xi(k)) + e(k) \quad (3.2.6)$$

As an identification or modelling problem the model parameters (n_y, n_u, d) as well as the approximation, $\hat{f}$, to f has to be determined that is parametrically parsimonious and satisfies various validation criteria of goodness of fit, such as d-step ahead output prediction (chapters 2 and 5), correlation tests (eg. autocorrelation residuals), and statistical tests such as the Chi-squared test to determine the approximating network bias [19]. This book only considers the class of feedforward supervised networks, where in addition to the input signals $\xi(k)$, the desired response $y(k)$ is available - the normal situation for input-output plant identification. Whereas unsupervised learning systems adapt the approximating network internal structure in an open loop manner, updating its parameters to represent the probability density of the input signal for instance; there is no performance feedback. Since the network has to learn inherent nonlinear plant input-output behaviours over the process operating envelope, the network's structure must contain appropriate nonlinearities. The learning capability of the network itself varies significantly according to the nonlinearities incorporated and how they are structured in the networks topology. However, they all share an important characteristic: they internally transform every training input into a higher dimensional space so that the network's output, $\hat{y}(k)$, can be made approximately linear to the transformed input. There are a variety of ways of interpreting the role the nonlinearities played in ANN. The most obvious is based on the relationship between the network output and its adjustable parameters, θ. This relationship is linear for lattice associative networks (LAN) such as the Cerebellum Model Articulation Controller (CMAC) and the B-spline [19], and linear optimisation techniques can be readily applied for weight adjustment. Whereas for the Multi-Layer Perceptron (MLP), the weight to output relationship is highly nonlinear, resulting in a cost function that is highly irregular, containing multiple minima and associated suboptimal weight solutions. Another useful way of describing ANN is based upon the type of generalisation by which the network transforms the input signal of measurement space. Networks can be generally classified as either globally or locally generalising. Generalisation is global if each network adaptable parameter or weight, θ, can potentially affect the network output for every point in the measurement or input space. Examples are the MLP and non-Gaussian Radial Basis Function (RBF) networks. The generalisation is local when only a few adjustable weights affect the network's output for each input; examples include CMAC, B-Splines as well as a class of fuzzy logic networks.

The type of neural network chosen depends on the application, eg. feature classification, image analysis, off-line controller design or on-line controller adaptation. The two extremes are feature/image analysis and on-line adaptive modelling/control, and these application areas place different (and sometimes conflicting) emphasis on the network's requirements. Parsimonious network structures are generally required in every application, but it may be that a network that is inherently difficult to train may be suitable for use with large input spaces, or that a network which learns quickly can only be applied to problems with a small or medium number of inputs.

The class of functions for which an important consideration of ANN is that of network approximation; given a set of L input vectors $\xi(k) \in R^n$ $(k = 1, 2, \ldots, L)$ and associated set of real numbers $y(k) \in R^n$, $(m = 1$ for simplicity and without loss of generality) the approximation problem is to find the function $f \in C(D)$ such that it satisfies the general interpolation condition:

$$f(\xi(k)) = y(k) \tag{3.2.7}$$

A large class of neural networks can be represented as:

$$S : \left\{ f \in C(D) : \ f(\xi) = \sum_{i=1}^{h} \theta_i^T \phi_i(\xi, \theta_i) \right\} \tag{3.2.8}$$

where θ_i is the (linear) output layer weight vector and ϕ_i is the ith hidden layer node or transfer function which may in turn have an internal set of parameters. With only mild restrictions on the type of nonlinearity incorporated in each node $\phi_i(.)$, a neural network can be shown to be able to approximate any continuous nonlinear function defined on a compact domain to an arbitrary accuracy. Hence they are called *universal approximators* (see [48, 107] for further details on ANN approximation theory). While these conditions ensure that the networks provide nonlinear functional approximation with arbitrary accuracy, this does not mean that they are best approximations[1]. To show that the sets S are also best approximants (or existence sets), it is only necessary to show that they are compact.

3.3 Multi-Layer Perceptrons

The MLP is currently the most widely used ANN, as it can be considered as a direct extension of the Perceptron and Adaline networks proposed in the sixties and the (re)invention of its gradient descent training rule called Back Propagation (BP) in 1986 by Rumelhart and McClelland [131] was a major factor in generating a resurgence of interest in these adaptive computational systems.

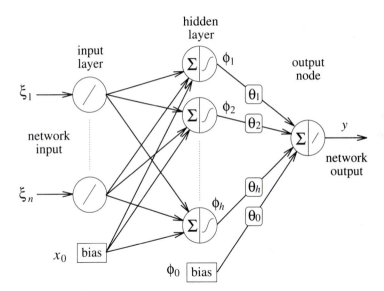

Figure 3.1: A typical 3-layer perceptron network with a linear output node.

The MLP may be represented by the network set

$$S_1 : \left\{ f \in C(D) : f(\xi) = \sum_{i=1}^{h} \theta_i \phi \left(\sum_{j=1}^{n} (\theta_{ij}\xi_j + \theta_{i0}) + \theta_0 \right) \right\} \qquad (3.3.1)$$

where θ_{i0} is a shift of threshold parameter in that it alters the value of the node's output when the information vector is identically zero, θ_i is the network's output, linear weight vector and θ_{ij} are the hidden layer weights and $\phi(\cdot)$ is the node activation function, which

---

[1]A best approximation is such that $\|f - \widehat{f}\|$ achieves its minimum value

are usually monotonic increasing, bounded and sigmoidal (or a squashing function with limits of 1 and 0 at $\pm\infty$, respectively). This is illustrated in Fig. 3.1. Typically the activation function is chosen as:

$$\phi(z) = \begin{cases} \dfrac{1}{(1 + \exp(-z))} & \in (0, 1) \\ \dfrac{1 - \exp(-2z)}{1 + \exp(-2z)} = \tanh(z) & \in (-1, 1) \end{cases} \tag{3.3.2}$$

where $z = \theta^T \xi + \theta_0$ is the activation of a node in the hidden layer. The nodes within an MLP are made up of an adaptive linear combiner and the sigmoidal like transfer functions which are termed *ridge* functions as the output is constant along the hyperplanes $\phi^T \theta_j + \theta_{j0} = c$ in their input space, as shown in Fig. 3.2. MLPs are good candidates for approximation and classification when the desired function or classification boundary can be concisely represented as similar ridge functions.

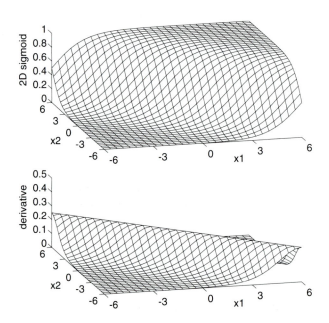

Figure 3.2: A 2-dimensional sigmoidal ridge function (top) and its derivative (bottom).

When each input ξ is quasi-normalised, the input to the jth network's node depends on the *projection* of the input vector ξ on the weight vector θ_j, *i.e.* the hidden layer's weight vectors represent feature or test vectors against which the input vectors are assessed or matched. Hence, MLPs are essentially row projection networks (see also section 3.6.1 on projection pursuit learning). The curse of dimensionality does not necessarily apply to

MLPs, therefore they are widely used in classification tasks with large feature spaces. Also MLPs construct global approximations (global generalisation) to input/output mappings, so they can extrapolate information to regions where the training data is sparse. However, because MLPs are globally generalising networks, the weights contain information about the whole of the input space, therefore when a network is trained in a certain region of the input space for a prolonged period of time, it can "forget" the information which relates to past experiences. Equally, this quality enables MLPs to be relatively robust to loss or damage of its weights. It is shown later that MLPs are not generally well-conditioned, therefore extensive training periods may be required, also the "squashing" of the output errors can lead to increased training times, when back propagation is used, as the hidden layers are relatively insensitive to network approximation errors.

In modelling and control applications, the input data maybe partially redundant, and MLPs model this relationship by constructing hyperplanes parallel to the redundant inputs with zero valued associated weights. The model's structure can incorporate redundant data efficiently, since as the input space is expanded by including new redundant input variables, no new nodes in the network need to be introduced. However, problems may occur in recognising this fact, especially using instantaneous gradient descent algorithms such as BP. The difficulty is in recognising this redundant information. MLPs are therefore appropriate for high dimensional functional modelling and classification tasks, when the training data has redundant inputs and the desired mapping can be approximated by a low number of ridge functions. The approximation capability of any ANN assumes that the network weights have been correctly assigned, although in reality this is an open question for MLPs. MLPs are generally unsuitable for modelling functions which have significant local variations, and establishing provable conditions for parameter (weight) convergence and stability for any learning law is very difficult as the resulting optimisation problem is nonlinear. Therefore in the following we concentrate on associative memory networks as being the most suitable for adaptive modelling and control.

3.4 Associative Memory Networks

Associative Memory Networks (AMNs) have many desirable properties for adaptive modelling and control: fast initial learning and long term convergence, and internal transparency which provides a qualitative interpretation of the knowledge stored in the network and local information storage which allows rapid retraining. Before describing their structure in detail however, some terminology must be introduced.

The *support* of a basis function $\phi_i(\xi)$ is the domain of the input space for which its output is non-zero:

$$\{\xi \in D : \phi_i(\xi) \neq 0\} \tag{3.4.1}$$

This is also known as its *receptive field*. A basis function has a *bounded support* when its support is (significantly) smaller than the input domain (this is also referred to as a compact support). The general topology of an AMN is the fixed (usually) nonlinear topology conserving mapping of the input space into a higher-dimensional hidden layer space, thus performing a pseudo-linearisation of the unknown function, which is followed by an adaptive linear mapping from the hidden to the output layer. This is illustrated in Fig. 3.3.

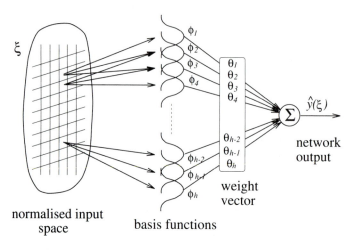

Figure 3.3: An associative memory network.

3.4.1 Radial basis functions

Radial Basis Function (RBF) networks were originally proposed as a method for modelling and interpolating data in high-dimensional spaces. RBFs are AMNs in which the basis functions $\phi_j(.)$ can be placed anywhere in the n-dimensional input space. A variety of learning algorithms have been used to train RBFs, such as an off-line orthogonal least squares algorithm which can be used to select the relevant basis functions or on-line clustering techniques which have been developed to modify their centres c_j. The network's weights θ_j can be found by direct matrix inversion or by recursive least squares methods or, for on-line adaptation, by using the instantaneous least mean square rules, as described in section 3.8. The topology of the RBF network can be summarised as $n - h - m$, where n is the network input dimension, h is the number of basis functions in the hidden layer and m is the network's output dimension, although this discussion shall concentrate on single output networks (without loss of generality). Each node in the hidden layer of the RBF network has a radially symmetric response profile around its centre c_j, and the overall network's output is given by:

$$y = \sum_{i=1}^{h} \theta_i \phi_i \left(\|\xi - c_i\|^2 \right)$$
(3.4.2)

where $\|.\|^2$ is the standard Euclidean norm and c_i $(= [c_{i,1}, \ldots, c_{i,n}])$ is the centre of the ith basis function ϕ_i. There are a variety of choices for the node nonlinearity $\phi_i(.)$, and typical examples are the Gaussian functions:

$$\phi_j(z) = \exp\left(\frac{-z^2}{\sigma_j}\right) \qquad (3.4.3)$$

which provide an interesting fuzzy interpretation of RBFs [159] as well as more formal links to orthogonal polynomial expansions of nonlinear functionals via Hermite polynomials. The thin-plate-spline function is another one that is widely used:

$$\phi_j(z) = z^2 \log(z) \qquad (3.4.4)$$

It can be seen that the type of basis function influences the network's modelling and learning abilities. For instance, when the basis functions have a significant amount of overlap (which generally occurs for the thin-plate-spline basis functions), the network becomes badly conditioned and learning is more difficult. However, the locally generalising Gaussian functions are not always appropriate for high-dimensional modelling problems and the globally generalising thin-plate-spline functions have been noted to have slightly better extrapolation abilities.

Supervised learning for generating hypersurfaces as models of the input/output process maybe ill-posed or overdetermined, since in general, the number of data points in the training set is much larger than the (unknown) degrees of freedom of the underlying physical process, and having as many basis functions as data points means that any inherent data noise will be modelled and the network's generalisation abilities will be poor. Ill-posed problems in approximation theory have been solved by *stabilising* the solution by imposing constraint conditions that incorporate a priori process such as functional smoothness. This maybe achieved by adding a constraint term to the optimisation performance function that incorporates the appropriate surface properties; this is termed *regularisation*. It is similar to simultaneous system parameter identification and model order determination [105] and to minimal neural network construction algorithms [14]. Smoothness between data points is not the only property required for approximation algorithms; it is also expected that the model be (almost) invariant under transformation of linear deformation of the data set (or equivalently, insensitive to incorrectly evaluated model parameters). An additional constraint which is a highly desirable property of neural networks is that of restricting functional approximators with localised receptive fields to those classes of basis functions which form *partitions of unity*, i.e. the unweighted sum $\sum_i \phi_i(z) \equiv 1$, $\forall z$. Since such networks can exactly reproduce the constant function $f \equiv 1$, they do not introduce false structure (or equivalently poor interpolation) between data points. Gaussian functions do not form a strict partition of unity whereas as the piecewise polynomial lattice-based B-splines do. However, any set of localised basis function can be trivially endowed with a partition of unity by defining a new set of basis functions as:

$$\phi_j^*(z) = \frac{\phi_j(z)}{\sum_{i=1}^{h} \phi_i(z)} \tag{3.4.5}$$

where $\phi_j^*(z)$ is the "new" j^{th} basis function and the set of new basis functions forms a partition of unity.

3.4.2   Lattice based associative memory networks

The output of a LAN is formed of a linear combination of overlapping basis functions which are evenly distributed in an n-dimensional subspace of R^n. Each of the basis functions is defined on a hyper-rectangular region in R^n which is known as its *support* or *receptive field*. Therefore the output of each basis function is non-zero only when the input lies in its receptive field. This feature endows the LANs with an ability to generalise locally: similar inputs are mapped onto nearby hyper-rectangular receptive fields which produce similar outputs, while dissimilar inputs are mapped onto distant hyper-rectangles and this produces independent outputs, as illustrated in Fig. 3.3.

Notice that the function approximation is only valid for a bounded input space because of the finite number of localised basis functions, although the function may be assumed to remain constant outside the bounded input space.

Model structure

The network output at time k, $\hat{y}(\xi(k))$, is computed by first transforming the input vector, $\xi(k)$, $(\in R^n)$ into a higher dimensional space (R^h) which is generated by the outputs of the h basis functions ϕ_i. The inner product of the transformed input vector, $\phi(\xi(k))$, (containing the outputs of the individual basis functions) with an h-dimensional adjustable weight vector, θ, is then calculated, so the overall network's output can be written as:

$$\hat{y}(\xi(k)) = \phi^T(\xi(k))\,\theta(k) \tag{3.4.6}$$

Each of the basis functions $\phi_i(\xi(k))$ has an associated support (region in the input space where its output is non-zero) and it is the shape of the basis functions and their supports as well as their distribution that give a LAN its distinguishing features. In the remainder of this section, the dependence on time k will be dropped from the discussion in order to simplify the notation.

Each multivariate basis function in an LAN is formed from the *projection* of n univariate basis functions onto each other and their shapes and sizes determine the form of $\phi_i(\xi)$. To be precise, let $\phi_{i,j}(\xi_j)$ represent the jth component of the ith multivariate basis function that is defined on the jth input axis. Then the multivariate basis function can be decomposed as:

$$\phi_i(\xi) = \phi_{i,1}(\xi_1)\widehat{*}\phi_{i,2}(\xi_2)\widehat{*}\cdots\widehat{*}\phi_{i,n}(\xi_n) \tag{3.4.7}$$

where $\hat{*}$ is a binary operator which is used to combine two basis functions[2], and two common examples are the product and minimum operator. Using either the product or the minimum operator means that the support of the multivariate basis function is an n-dimensional hyper-rectangle where the length of each side is equal to the support of the respective univariate basis function. The type of operator *does* influence the shape of the multivariate basis function and the product operator is generally used as the output surface is generally smoother. It is also recommended to normalise the basis functions in order to ensure a partition of unity (see Eq.(3.4.5)). There are many ways for choosing the univariate basis functions (generalised B-splines, compact Gaussians etc., see Fig. 4.2) and the network's output surface will reflect their form. In Fig. 3.4, two triangular univariate basis functions are combined using the product operator.

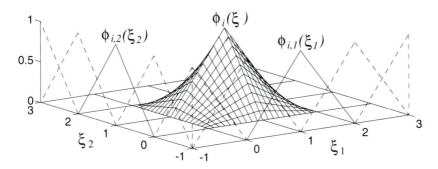

Figure 3.4: The generation of a multivariate basis function $\phi_i(\xi)$.

Each of the supports of the multivariate basis functions is n-dimensional hyper-rectangles and is positioned on a user-defined n-dimensional *lattice* or grid. A set of knots is specified on each of the input axes and the overall lattice is generated from their axis orthogonal projections in n-dimensional space, as illustrated in Fig. 3.4. The univariate basis functions are defined on the appropriate univariate knot sets, and the distance between two adjacent knots is, roughly speaking, the width of a univariate basis function. More specifically, the ith input axis is partitioned into r_i neighbouring, non-overlapping intervals by a set of $(r_i + 1)$ knots, λ_i. The first and last knot are termed *exterior* knots as they are positioned at the minimum and maximum values of ξ_i respectively. The remaining $r_i - 1$ values are called *interior* knots representing the position of the end of one interval and the beginning of the next. Each univariate basis function is non-zero only over a small number of adjacent intervals (ρ_i), therefore the position of the knots defines the size of the univariate supports and determines how the network generalises.

In general, the intervals on each axis are equi-spaced in order to preserve a uniform

---

[2]This operator is "borrowed" from the fuzzy logic literature where it implements a fuzzy intersection using a $t - norm$.

resolution of the network response in R^n. However, when *a priori* knowledge is available, the knots can be arranged so that the network's resolution is higher in the parts of input space where the function varies significantly, and a coarse representation is formed in the remaining areas.

The basis functions in a LAN are *evenly* distributed on the lattice. This is equivalent to say that each input activates κ basis functions. This observation allows the supports of the basis functions to be grouped into n-dimensional *overlays* where an overlay is a *coarse* lattice which is formed from adjacent, non-overlapping supports. When an input is presented to the network, one and only one basis function on each overlay is active (because the supports are adjacent and non-overlapping), and this allows identical computations to be performed on each of the κ overlays independently and also allows the weight vector to be stored in κ smaller packets. Each or the coarse lattice overlays is then placed on the original input lattice such that it is displaced relative to all the other overlays, and this uniquely determines the position of all of the basis function's supports, as illustrated in Fig. 3.5. The ith overlay is displaced by an amount d_i (an n-dimensional vector) relative

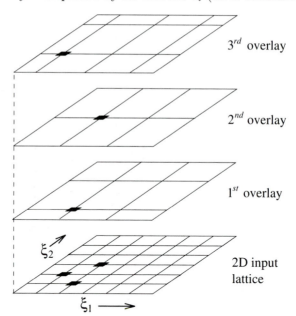

Figure 3.5: A set of $\kappa = 3$ overlays which cover the original input lattice.

to the origin of the original input lattice, where $1 \leq d_{i,j} \leq \rho_i$.

Based on the lattice defined by Λ and the basis function distribution defined by D, the number of basis functions h in the "hidden layer" can be calculated from:

$$h = \sum_{j=1}^{\kappa} \prod_{i=1}^{n} \left(\left\lceil \frac{r_i + 1 - d_{i,j}}{\rho_i} \right\rceil + 1 \right) \tag{3.4.8}$$

CMAC

The CMAC network was originally proposed [1] as a model of the neuro-physiological functioning of the mammalian cerebellum. However, it aims to model the *functioning* of the neuronal substrate rather than the individual neurons directly. This ensures that desirable properties such as localised representation and learning are a feature of the algorithm, rather than hoping that they will emerge during training.

The CMAC is a LAN where the size of the basis functions' supports and the size of the network's response region are both determined by the user-supplied *generalisation parameter*, ρ. For each input exactly ρ basis functions are active ($\kappa = \rho$) and each support is a hypercube of volume ρ^n. Therefore choosing a large generalisation parameter means that a lot of basis functions contribute to the output and the support of each basis function is relatively big (with respect to the lattice). Another unique feature of the CMAC is its algorithm for sparsely coding the original input lattice with a set of much coarser overlays. As was described in the previous section, the basis functions' supports can be grouped into overlays which are then displaced relative to each other on the input lattice. The CMAC forms ρ overlays consisting of hyper-cubic supports of volume ρ^n and the original method for distributing the overlays was to position their origin along the main diagonal of the much finer input lattice. The displacement vector for the ith overlay is:

$$d_i = (i, i, \ldots, i), \qquad \text{for } i = 1, 2, \ldots, \rho \qquad (3.4.9)$$

as shown in Fig. 3.6, where the origin of each overlay occurs at the lattice point denoted by a dark square. These features ensure that the computational cost of retrieving the

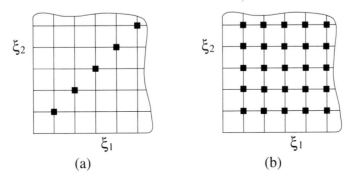

Figure 3.6: The CMAC (a) and B-splines (b) overlay structures for $\rho = 5$.

information stored by the CMAC is *linearly* dependent on n and that the number of basis functions used by the network is *significantly* smaller than the number of cells in the original input lattice. However, this basis function distribution method is non-even in the sense that the size and shape of the overall network's response region is strongly dependent on the position of the input. This can be overcome by using more advanced overlay displacement algorithms [2, 126] and these are summarised in [19] which also provides an example that shows how the approximation ability of the network is significantly improved when the generalised overlay displacement vectors are used.

The original CMAC uses binary basis functions which are on ($= 1$) when an input lies in their support and off ($= 0$) otherwise. This means that the network has a *piecewise constant* output: whenever the input moves between cells the network's response is discontinuous but within a lattice cell its response is constant. This is equivalent to a look-up table (and indeed for $\rho = 1$ the CMAC is a look-up table), and for $\rho > 1$, the CMAC is a form of look-up table which locally generalises. Higher order, smoother basis functions can be used in the network to generate a continuous output, and an example is shown in Fig. 3.7 which compares binary and triangular basis functions [2, 19, 84].

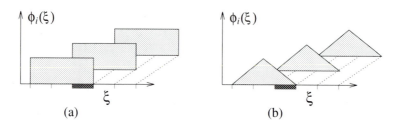

Figure 3.7: Binary (a) and piecewise linear (b) basis functions used in the CMAC.

Despite the binary CMAC being similar to a look-up table, it should be noted that because it uses far fewer parameters, it is not as flexible. An extensive investigation has been made into which look-up tables a CMAC can and cannot reproduce in [19, 17], and the main result is that the generalisation parameter ρ implicitly specifies the size of the regions (with respect to the lattice) where the CMAC forms an *additive*[3] model. Within these additive regions, the CMAC cannot store any multiplicative information of the form $\xi_1\xi_2$, but because these additive regions are relatively small and evenly distributed on the lattice, the network can successfully approximate most smooth functions.

B-splines

B-splines were originally developed for use in geometrical modelling and in graphical applications. In its most basic form, a B-spline surface is formed from a linear combination of piecewise polynomial basis functions defined on a lattice. It is therefore obvious that they can be regarded as LANs, and the B-spline LANs can also be used to represent fuzzy, linguistic rules, as described in chapter 4. This latter relationship is particularly significant since it offers both the opportunity to incorporate a priori common sense, expert knowledge in the form of linguistic rules into the network, but also provides a mecha-

---

[3]An n-dimensional additive model is of the form $\widehat{y}(\xi) = \sum_{i=1}^{n} \widehat{y}_i(\xi_i)$, where $\widehat{y}_i(.)$ is a univariate function.

nism for network transparency through the linguistic rules that represent the functional relationship.

The shape of the univariate B-spline basis functions (and also the width of their support) is given when the user specifies their *order*, k_i. The univariate basis functions on the ith axis are then piecewise polynomials of order k_i and are ($\rho_i = k_i$) intervals wide, as shown in Fig. 3.8. The basis functions' output can also be calculated using a simple

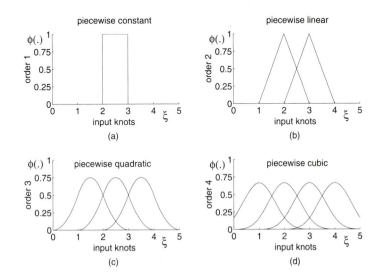

Figure 3.8: B-spline univariate field shapes of orders $k = 1, \ldots, 4$.

and stable recurrence relationship as follows:

$$\phi_{i,k_i,j}(\xi_i) = \left(\frac{\xi_i - \lambda_{i,j-k_i}}{\lambda_{i,j-1} - \lambda_{i,j-k_i}} \right) \phi_{i,k_i-1,j-1}(\xi_i) + \tag{3.4.10}$$

$$\left(\frac{\lambda_{i,j} - \xi_i}{\lambda_{i,j} - \lambda_{i,j-k_i+1}} \right) \phi_{i,k_i-1,j}(\xi_i) \tag{3.4.11}$$

$$\phi_{i,1,j}(\xi_i) = \begin{cases} 1 & \text{if } \xi_i \in I_{i,j-1}; \\ 0 & \text{otherwise} \end{cases} \tag{3.4.12}$$

where $I_{i,j}$ is the jth interval $[\lambda_{i,j}, \lambda_{i,j+1})$ on the ith axis, with the last interval being closed at both ends. This relationship illustrates how a first order B-spline is a binary function while a second-order B-spline is a piecewise linear function, and choosing the order of a basis function is equivalent to specifying the smoothness of the network's output.

The order of univariate basis functions implicitly specifies the number of basis functions that contribute to the network's output $\kappa = \prod_{i=1}^{n} k_i$ as well as influencing the total number of basis functions $h = \prod_{i=1}^{n}(r_i + k_i)$. Both expressions grow as an exponential function of the number of inputs and may seen to limit the network's applicability to only small modelling problems. However, new network construction algorithms are being developed, see section 4.3.2, which explicitly model any redundancy in the desired function and as such these techniques are applicable to medium (≤ 20) dimensional input spaces as well. The κ overlays are distributed at every available point in the lattice, as illustrated in Fig. 3.7, which illustrates how the CMAC can be regarded as a sparsely coded B-spline network that uses a set of generalised (dilated [84]) basis functions.

3.5   Polynomial Neural Networks

Polynomial Neural Networks (PNNs) are single-layer devices which use higher-order correlations of input components to perform global nonlinear mappings. Examples of PNNs include sigma-pi and functional layer networks, which attempt to approximate a given multivariate function $f : \Re^n \to \Re$ by:

$$\widehat{f}(\xi_1, \ldots, \xi_n) \;=\; \theta_0 + \sum_{i=1}^{n} \theta_i \xi_i + \sum_{i_1=1}^{n} \sum_{i_2=i_1}^{n} \theta_{i_1,i_2} \xi_{i_1} \xi_{i_2} + \cdots + \tag{3.5.1}$$

$$\sum_{i_1=1}^{n} \sum_{i_2=i_1}^{n} \cdots \sum_{i_k=i_{k-1}}^{n} \theta_{i_1,i_2,\ldots,i_k} \xi_{i_1} \xi_{i_2} \cdots \xi_{i_k} \tag{3.5.2}$$

$$=\; \theta^T \phi(\xi) \tag{3.5.3}$$

where θ is the concatenated weight vector, ϕ is the vector of basis functions formed from the polynomial input terms and k is the order to the polynomial expansion. This is similar to the well-known Gabor-Kolmogorov polynomial expansion which has the important property of being *linear* in its adjustable parameters which ensures that learning is quite fast. But the number of weights required to accommodate all the high order correlations is:

$$h = \frac{(n + k - 1)!}{n!(k - 1)!} \tag{3.5.4}$$

and in practice $k \geq 3$ which leads to enormous hidden layers, even for moderately sized input spaces.

Automatic techniques have been developed which can reduce the number of terms in the polynomial expansion by selecting the most relevant terms [22]. Also hierarchical models have been proposed which perform a low order polynomial transformation on each layer and successively increase the order of the approximation as more layers are included [70]. By investigating different representations for the polynomials, *ridge* polynomial networks have been proposed which can be considered as a special case of the more general projection pursuit algorithm [139].

3.6 The Curse of Dimensionality

Generally associative memory networks (AMN) suffer from the curse of dimensionality, in that the memory storage (and computational cost) increases exponentially with input dimension n. For many practical dynamical systems (eg. a helicopter) the input space maybe > 18, therefore to exploit the algorithmic power of AMN for high dimensional input spaces it is necessary to develop off-line constructional algorithms that structure the network appropriately by exploiting any *redundancy* in the unknown relationship.

To illustrate this problem, consider trying to model the following desired function which is defined on a four dimensional unit hypercube, to a resolution of 0.1 on each axis:

$$y = (\xi_1 + 10\xi_2)^2 + 5(\xi_3 - \xi_4)^2 + (\xi_2 - 2\xi_3)^4 + 10(\xi_1 - \xi_4)^4 \tag{3.6.1}$$

Then a minimum of 10^4 observations are required and as the input space dimension grows, this becomes infeasible. However, when *four* AMN subnetworks were used to model the desired function as follows:

$$\hat{y} = \hat{f}_1(\xi_1, \xi_2) + \hat{f}_2(\xi_3, \xi_4) + \hat{f}_3(\xi_2, \xi_3) + \hat{f}_4(\xi_1, \xi_4) \tag{3.6.2}$$

where $\hat{f}_1, \hat{f}_2, \hat{f}_3, \hat{f}_4$ are 2-dimensional AMN's, each subnetwork requires only 10^2 observations and the overall network would need only 400 training pairs. This is because the *global, additive* redundancy in the desired function is being explicitly modelled by the network and any unnecessary cross product terms are not included in its formulation. This additive subnetwork scheme is illustrated in Fig. 3.9.

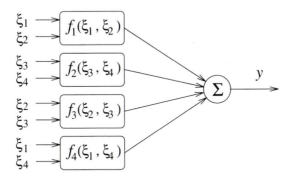

Figure 3.9: Additive network decomposition.

In general, a multivariate function f has an ANalysis Of VAriance (ANOVA) expansion which is given by:

$$f(\xi) = f_0 + \sum_{i=1}^{n} f_i(\xi_i) + \sum_{i=1}^{n}\sum_{j=i+1}^{n} f_{i,j}(\xi_i, \xi_j) + \cdots + f_{1,2,\ldots,n}(\xi) \tag{3.6.3}$$

and each additive subfunction f_* could be implemented by an ANN subnetwork. Then if the total size of all the subnetworks was smaller than a full n-dimensional network, the ANOVA structure would be a more *parsimonious* representation. It is important to notice that the ANOVA representation only discovers the *additive* structure between smaller dimensional subnetworks, it does not specify how the individual subnetworks should be represented.

A variety of constructional algorithms have been proposed in the literature according to whether they exploit *additive* redundancy in $f(\phi)$, for example:

- Adaptive Spline Modelling of Observational Data (ASMOD) [78],

or *additive and subnetwork* redundancy:

- Adaptive B-spline Basis function Modelling of Observational Data (ABBMOD) [14],

- Tree structured Adaptive Approximation (TAA) [135],

- Multivariate Adaptive Regression Splines (MARS) [52].

Algorithms which exploit additive and local redundancy also search for local regions in the input space where there is no dependency between the output and certain combinations of the input, however this strategy causes extra computation in generating the input/output approximation.

In many of the above statistical based algorithms, the most popular solution is divide and conquer (as in the MARS algorithm [52]) which utilise algorithms that fit simple, low order surfaces to data by dividing the input space into nested sequences of regions. This approach tends to increase the variance while reducing the model's bias, which is due to data outliers contributing most to variance evaluation. MARS copes with the problem by utilising hard split algorithms and fitting piecewise constant or linear functions in each local region, so as to minimise variance at increased bias. A further reduction in variance can be achieved using soft splits instead, which allows neighbouring data to influence different regions (through appropriate overlap). These regional boundaries may also be parameterised surfaces that are optimised by an offline learning algorithm to generate the most parsimonious representation. Jacobs and Jordan [74] have achieved this via a hierarchical mixture modelling method that also minimises bias by using locally linear functionals. The (soft) splits are formed along arbitrarily oriented hyperplanes in the input space so as to reduce the interaction among the inputs as well as the sensitivity to the coordinates selected to endorse the data. This overcomes one of the main limitations of the MARS algorithm which is restricted to forming *axis orthogonal* splits.

3.6.1 Projection pursuit learning

A more direct approach to overcoming the curse of dimensionality in neural networks is found in the statistics literature through projection pursuit regression, which assumes that the data can be modelled by networks of the form:

$$y = \theta_0 + \sum_{i=1}^{h} \theta_i \phi_i(\alpha_i^T \xi) + \epsilon \qquad (3.6.4)$$

where ϵ is the approximation error satisfying $E(\epsilon) = 0$, $\alpha_i^T \xi$ is a 1-dimensional projection of ϕ on α_i (where $\|\alpha_i\|^2 = 1$) and ϕ_i is an arbitrary univariate function with $E(\phi_i(.)) = 0$, $E(\phi_i^2(.)) = 1$ (i.e. all the functions are zero-mean and have the *same* power). Smoothing is always 1-dimensional with the neighbourhoods being infinitely large in directions orthogonal to the projection; this is based on the assumption that the surface is constant in these directions. There are some problems in interpretating the resulting approximation for $h > 1$, but some special cases are the semi-parametric additive models in the independent variables ξ_i, where each vector α_i has a unity value in one entry and is identically zero otherwise. This representation retains the linear regression properties of separating variable effects and the output, with the functions ϕ_i akin to the coefficients in linear regression. The approach can be used with the full range of multiple regression techniques in which variables maybe transformed parametrically by power transformations (such as log, square root etc.), generate sets of variables such as orthogonal polynomials which after fitting the function maybe reconstructed. These ideas have evolved into the two layer Projection Pursuit Learning algorithm (PPL). It is called PPL since it interprets high dimensional data through selected 1-dimensional projections and all the parameters (the weights and projection vectors) are trained to minimise the output mean squared error using Quasi-Newton techniques. As in an MLP, the PPL network forms projections of the data in directions determined by the interconnection weights, but generally the activation functions are more flexible than the standard sigmoid used in the MLP. The PPL procedure uses a batch, backfitting learning algorithm to iteratively minimise the loss function over the training data until some convergence criteria is satisfied.

3.7 Supervised Learning

New training algorithms and their associated convergence and stability proofs are behind the recent revival of interest in neurocomputing. The inability of the original perceptron training algorithm to be extended to multi-layer networks caused a sharp decline in funding for neural network research, although in the mid-eighties the so called δ or back propagation rule was developed [131] for training these multi-layer perceptrons leading to a revival in this field. The "new" learning algorithm was simply a nonlinear generalisation of instantaneous gradient (of steepest) descent. However, the training rule's biological interpretations were stressed as well as an efficient procedure for back-propagating the network's error performance through several layers. Multi-layer perceptrons trained using back propagation have been applied to many different problems (prediction, estimation, classification and control) in the late eighties and it was gradually realised that there are many different second-order optimisation algorithms which can be used to estimate the unknown parameters. These higher order training rules generally result in improved parameter convergence and generalisation characteristics, although they have no biological interpretation.

Many of the so-called biologically inspired training algorithms have been developed and applied in the fields of signal processing, adaptive control, optimisation theory as well as being used to train neural networks [66]. A lot of theory has been developed in these fields which can predict how the learning rules behave and is therefore directly applicable to neural systems. Also the training rules which are nonlinear generalisations of conventional linear algorithms can be studied by linearising about the network's current state [48]. By using techniques developed in related research fields, the mystique surrounding the "black box" neural network can be gradually removed.

This book only considers the set of *supervised* training algorithms, where the plant's desired response is available as an external signal. Unsupervised learning systems adapt the network's interval structure in an "open loop" fashion, where the parameters are trained using only the input signal received; there is no performance feedback. For many plant modelling and control tasks, some form of desired signal is available and is generally produced via a reference model as in conventional adaptive control which specifies how the system should behave.

It is assumed that there exists a set of input/output training pairs $\{\xi(k), y(k)\}_{k=1}^{L}$, where $y(k)$ is the desired output for given input $\xi(k) \in R^n$. Supervised training algorithms can be divided into two distinct groups: off-line, batch or epoch training in which the complete data set (or a subset) is used to train the network, and on-line, instantaneous training by which the network is adapted as each input/output sample is taken. Off-line algorithms are frequently used to compute the internal structure (node or basis centres, generalisation width, network functional construction etc) of ANN as well as its adjustable weights, whereas on-line algorithms are usually reserved for evaluating weights alone. On-line modelling and control via ANN is essential for dynamic processes such as gas turbines, missiles, high incidence aircraft, large scale chemical (STR etc), where the dynamics vary nonlinearly and with time. Associative memory networks (AMN) such as the CMAC, the B-spline, Gaussian Radial Basis Function (RBF) and a certain class of adaptive fuzzy networks have many desirable features for *on-line* modelling and control, since their network structure can be simply expressed as:

$$\widehat{y}(k) = \phi^T(k)\,\theta(k) \qquad (3.7.1)$$

where:

$$\phi(k) = \phi(\xi(k)) \subseteq [0,1]^h \quad (h \gg n) \qquad (3.7.2)$$

is a fixed set of basis functions in which only a small number, ρ are activated for each input, so the network's output is locally stored across a few weights. Linearity in the adjustable weights leads to simple training rules for which rate of convergence and stability results can be derived.

3.7.1   System of linear equations

Given a data set $\{\xi(k), y(k)\}_{k=1}^{L}$ and a model specified by Eq.(3.7.1), the parameter identification problem can be expressed as:

Solve the following set of linear equations for the unknown set of parameters
θ:

$$A\theta = y \qquad (3.7.3)$$

where A is an $(L \times h)$ matrix whose elements are given by $a_{ij} = \phi_j(i)$ and y
is an L-dimensional column vector composed of the desired outputs.

The solution to this problem can be given in terms of the *generalised inverse*, and the
"optimal" weight vector $\hat{\theta}$ is:

$$\hat{\theta} = A^{\dagger}y \qquad (3.7.4)$$

Depending on the form of the matrix A, the generalised inverse reduces to well-known
special cases. For instance, when the data are inconsistent but spanning, the generalised
inverse is given by:

$$A^{\dagger} = (A^{T}A)^{-1}A^{T} \qquad (3.7.5)$$

which is simply the Moore-Penrose pseudo-inverse. Alternatively, when the data are
singular and consistent, the optimal weight vector is calculated according to:

$$A^{\dagger} = A^{T}(AA^{T})^{-1} \qquad (3.7.6)$$

although the most general formulation is in terms of its Singular Valued Decomposition,
which allows the generalised inverse to be written as:

$$A^{\dagger} = V \begin{bmatrix} \Lambda^{-1} & 0 \\ 0 & 0 \end{bmatrix} U^{T} \qquad (3.7.7)$$

where V and U are unitary matrices composed of the right and left singular vectors,
respectively, and Λ is a diagonal matrix composed of the (ordered) singular values of A
[66].

This expression may initially seem a little complex but many of the concepts that
appear in supervised learning (i.e. minimum norm solutions, mean square solutions) also
occur in linear least squares data fitting problems, as will be highlighted in this chapter.
As long as the search space is the same for each of these learning rules and the step sizes
are calculated in a consistent manner, most training algorithms converge to the same
optimal solution.

3.7.2   Performance surface

The network's output error $\epsilon_y(k) = y(k) - \hat{y}(k)$, for an available (*i.e.* this is a supervised learning network) desired network output $y(k)$, gives an instantaneous measure of current model performance. Learning laws which adjust weight vector θ, must minimise prespecified cost functions that utilise all available training data, such as:

$$J = \begin{cases} E(|\epsilon_y(k)|) \\ E(\epsilon_y^2(k)) \\ \max_k |\epsilon_y(k)| \end{cases} \tag{3.7.8}$$

Unless there exists a weight vector which models the desired function exactly, then the optimal weight vectors for each of these cost functions are *different* as the emphasis placed on each of the (nonzero) instantaneous errors is different. The cost function also influences the complexity of the learning rule, for instance very complex learning rules are needed to minimise the unitary cost function $\max_k |\epsilon_y(k)|$. The choice of cost function also determines the type of learning rule and its computational complexity, as well as the final model generated. For instance, the MSE cost function is typically adopted because it gives satisfactory performance, and the resulting learning rules are simple to implement. The final models produced are not always the "best", rather the MSE is generally a good engineering choice.

The MSE (and the other cost functions) can be interpreted as a *performance surface* in $(h + 1)$ dimensional space, where the scalar cost is expressed as a function of the h weights. To generate the performance surface, the MSE for a static weight vector θ is evaluated:

$$J = E(\epsilon_y^2(k)) \tag{3.7.9}$$

where the expectation operator $E(.)$ is the discrete averaged sum. The network output, given in Eq.(3.7.1) is linearly dependent on the weight vector θ, and so the MSE cost function forms a quadratic bowl in the weight space. Substituting Eq.(3.7.1) into Eq.(3.7.9) gives after simplification

$$J = E(y^2(k)) + \theta^T R\theta - 2p^T\theta \tag{3.7.10}$$

where the real, symmetric, semi-positive definite *autocorrelation matrix* R is defined as:

$$
\begin{aligned}
R &= E(\phi(k)\phi^T(k)) = \frac{1}{L}AA^T \\
&= \begin{bmatrix}
E(\phi_1^2(k)) & E(\phi_1(k)\phi_2(k)) & \cdots & E(\phi_1(k)\phi_h(k)) \\
E(\phi_2(k)\phi_1(k)) & E(\phi_2^2(k)) & \cdots & E(\phi_2(k)\phi_h(k)) \\
\vdots & \vdots & & \vdots \\
E(\phi_h(k)\phi_1(k)) & E(\phi_h(k)\phi_2(k)) & \cdots & E(\phi_h^2(k))
\end{bmatrix}
\end{aligned}
$$

and each element represents the power of interaction between two (not necessary distinct) transformed inputs. The *cross correlation* vector, p, is defined as:

$$p = E(y(k)\phi^T(k)) = \frac{1}{L}A^T y \tag{3.7.11}$$

$$= \begin{bmatrix} E(y(k)\phi_1(k)) \\ E(y(k)\phi_2(k)) \\ \vdots \\ E(y(k)\phi_n(k)) \end{bmatrix} \qquad (3.7.12)$$

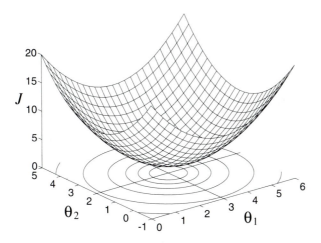

Figure 3.10: Network performance surface in two-dimensional weight space.

When the autocorrelation matrix, R, is non-singular, the optimal MSE weight vector, $\hat{\theta}$, can be found by differentiating Eq.(3.7.10) with respect to θ, equating to zero and solving for θ as

$$\hat{\theta} = R^{-1}p \qquad (3.7.13)$$

It can be clearly seen that this is just the Moore-Penrose pseudo-inverse solution for inconsistent, spanning training data (see Eq.(3.7.5)). In fact, minimising the MSE performance function and solving a set of linear equations is equivalent whenever the (modified) matrix is symmetric, and in this case the autocorrelation matrix is symmetric.

As well as investigating the structure of the autocorrelation matrix, it is worthwhile looking at the form of the cross-correlation vector as it can indicate the type of functions that a network may have difficulty in modelling. In particular, a non-zero desired function which produces an identically zero cross-correlation vector is called an *orthogonal* function as the optimal weight vector and hence the network's output is also identically zero [19]. This occurs when the expected value of the desired function multiplied by each basis function is zero *i.e.*

$$p_i = E(y(k)\phi_i(k)) = 0 \qquad \forall i \qquad (3.7.14)$$

The minimum MSE, J_{min}, which occurs when $\theta = \hat{\theta}$, can now be found by substituting

$\widehat{\theta}$ into Eq.(3.7.10), and noting that R is symmetric,

$$J_{min} = E(y^2(k)) - p^T\widehat{\theta} \tag{3.7.15}$$

Then defining $\epsilon_\theta(k) = \widehat{\theta} - \theta(k)$ as the *current* weight vector error, substituting Eq.(3.7.13) and Eq.(3.7.15) into Eq.(3.7.10) gives on simplification:

$$J = J_{min} + \epsilon_\theta^T(k)R\epsilon_\theta(k) \tag{3.7.16}$$

From this expression it can be clearly seen that the performance function forms a quadratic bowl (as R is positive definite) around the point $\theta = \widehat{\theta}$. When $\epsilon_\theta(k)$ lies in the null space of R, or is zero, $J = J_{min}$, otherwise any error in the weight vector causes an increase in J. Thus the autocorrelation matrix R determines the shape of the performance surface and this is discussed further in the following sections.

Normal form

Performance surface properties such as steepness and orientation are directly related to the structure of the autocorrelation matrix R, which can be expressed in its normal form

$$R = Q\Lambda Q^{-1} = Q\Lambda Q^T \tag{3.7.17}$$

where Λ is an $(h \times h)$ diagonal matrix composed of the (non-negative) eigenvalues of R and Q is a unitary matrix whose columns are the orthonormal eigenvectors of R. The MSE Eq.(3.7.10) can be expressed in terms of the eigenvalues of R by combining Eqs (3.7.16) and (3.7.17) as

$$J = J_{min} + v^T\Lambda v = J_{min} + \sum_{i=1}^{h} v_i^2\lambda_i \tag{3.7.18}$$

where $v = Q^T\epsilon_\theta$. Hence J is a hyper-ellipsoid in v-space, with a minimum occurring at the origin and with p lines which are perpendicular to all the contours of J. These lines are known as the *principal axes* of the ellipse. It can easily be shown that these principal axes are simply the eigenvectors of the autocorrelation matrix. Also as Q is a unitary matrix, then v is simply a rotated version of ϵ_θ in the θ-space (weight space), with an origin at $\widehat{\theta}$.

The eigenvalues of R, which are contained in the diagonal matrix Λ, represent the second derivative of J, along any of the *principal axes* as:

$$\frac{\partial^2 J}{\partial v_i^2} = 2\lambda_i \qquad \forall i \tag{3.7.19}$$

which means that the second partial derivatives of J are simply twice the corresponding eigenvalues of the autocorrelation matrix. The second derivative contains the *curvature*

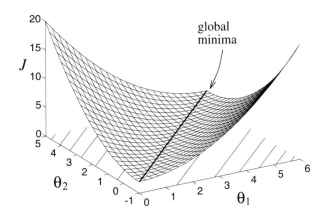

Figure 3.11: Performance surface for a singular network.

information about the cost function, and so if the eigenvalues of the autocorrelation matrix are widely spread then this is reflected in the relative steepnesses of the performance surface along the principal axes. In particular if $\lambda_i = 0$, the cost functional is flat along the corresponding principal axis, as shown in Fig. 3.11.

The spread of eigenvalues of R, can be measured through the *condition* number $C(R)$, defined as the maximum eigenvalue divided by smallest positive eigenvalue, (R has non-negative eigenvalues due to its structure). It will be shown in the following that the rate of convergence of a weight vector (and hence learning) that is adapted using gradient based methods depends on $C(R)$.

3.7.3 Batch gradient based methods

Gradient based learning algorithms update the weight vector parallel to the negative gradient of the performance surface such that the MSE is reduced. The gradient of J is denoted by ∇ and can be obtained from differentiating Eq.(3.7.10) with respect to θ, giving:

$$\nabla = \frac{\partial J}{\partial \theta} \tag{3.7.20}$$

$$= 2R\theta - 2p = -2R\epsilon_\theta \tag{3.7.21}$$

Hence ∇ depends only on the autocorrelation matrix R and the current weight vector error. Also at $\theta = \hat{\theta} = R^{-1}p$, and $\nabla = 0$. An important insight can be gained by decoupling the performance surface and transforming to the v-space as the gradient can then be expressed as (see Eq.(3.7.18)):

$$\frac{\partial J}{\partial v} = 2\Lambda v \tag{3.7.22}$$

The gradient of J, with respect to v_i, only depends on the ith eigenvalue, λ_i, of R, and v_i, and does not depend on (λ_j, v_j) for $i \neq j$.

Gradient descent

Gradient descent learning rules update the weight vector in proportion to the negative gradient of J. This produces a learning rule of the form:

$$\begin{align}
\Delta\theta(k) &= -\delta(R\theta(k) - p) \tag{3.7.23} \\
&= \delta R\epsilon_\theta(k) \tag{3.7.24}
\end{align}$$

where $\Delta\theta(k) = \theta(k+1) - \theta(k)$, δ is the scalar learning rate. The learning rate determines the rate of convergence and the stability of this adaptation rule.

The gradient descent learning algorithm moves the weight vector perpendicularly to the contours in the cost function, and hence the structure of the autocorrelation matrix, which determines the shape of the performance surface, also influences the rate of weight convergence. This is illustrated in Fig. 3.12, for $C(R) = 3$ and $\delta = 0.5$. The slow convergence along the flattest principal axis (corresponding to the smallest eigenvalue) is due to the learning rate being selected to stabilise the fastest mode which slowing down convergence of the other modes.

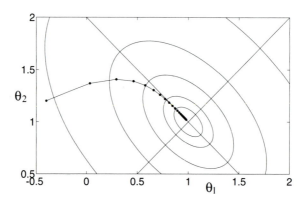

Figure 3.12: Gradient descent on a performance surface with $C(R) = 3$ and $\hat{\theta} = (1,1)^T$.

Steepest descent

The steepest descent learning rule (recursively) calculates the one-step-ahead optimal step size and moves parallel to the negative gradient this amount. Therefore, it is more complex than the standard gradient descent rule but it generally converges at a faster rate, and the algorithm is expressed as:

$$r(k) \quad = \quad R\theta(k) - p \qquad\qquad (3.7.25)$$

$$\delta(k) \quad = \quad \frac{r^T(k)r(k)}{r^T(k)Rr(k)} \qquad\qquad (3.7.26)$$

$$\Delta\theta(k) \quad = \quad -\delta(k)\,r(k) \qquad\qquad (3.7.27)$$

However, the residual vector $r(k)$ can be calculated in an iterative fashion as can be seen from premultiplying both sides of Eq.(3.7.27) by R and adding p, which gives:

$$\Delta r(k) = -\delta(k)Rr(k) \qquad\qquad (3.7.28)$$

Eq.(3.7.26) must be used to calculate $r(0)$, although thereafter the recursive expression (3.7.28) can be used and a saving of one matrix, vector product occurs as the term $r(k)$ which appears in Eq.(3.7.26) and Eq.(3.7.28) only has to be calculated once. However, any numerical roundoff errors may cause the iterative solution to drift away from the true value, and every so often the true value must be calculated using Eq.(3.7.26).

Rate of convergence

The dynamic behaviour of the gradient descent law for AMN can be investigated by subtracting $\hat{\theta}$ from Eq.(3.7.24) to give:

$$\epsilon_\theta(k+1) = (I - \delta R)\epsilon_\theta(k) \qquad\qquad (3.7.29)$$

Premultiplying by Q^T gives:

$$v(k+1) = (I - \delta\Lambda)v(k) \qquad\qquad (3.7.30)$$

or rewritten in its decoupled form (p-independent equations)

$$v_i(k+1) = (1 - \delta\lambda_i)v_i(k) \qquad \forall i \qquad\qquad (3.7.31)$$

which have the closed form solutions:

$$v_i(k+1) = (1 - \delta\lambda_i)^k v_i(1) \qquad \forall i \qquad\qquad (3.7.32)$$

where $v(1)$ is the (rotated) initial error in the weight vector.

For stable learning each term in these first order difference equations must have a modulus of less than or equal to one, otherwise any initial error will grow without bound. Also, when

$$|1 - \delta\lambda_i| < 1 \qquad \forall i \tag{3.7.33}$$

the initial errors in the weight vector converge to zero exponentially fast. However, the learning rate δ must be selected such that the fastest mode is stable:

$$0 < \delta < \frac{2}{\lambda_{\max}} \tag{3.7.34}$$

and generally the learning rate is selected to be $\delta = 1/\lambda_{\max}$, but this can seriously limit the rate of convergence of the other modes. In particular, let λ_i represent the smallest positive eigenvalue, then its dynamical behaviour is given by:

$$v_i(k + 1) = (1 - 1/C(R))^k v_i(1) \tag{3.7.35}$$

where $C(R)$ is known as the *condition* of the autocorrelation matrix and is defined by $C(R) = \lambda_{\max}/\lambda_{\min}$. When the network is *well-condition*, its condition number is close to unity and the weight vector converges very fast as the performance function is bowl shaped. However, the rate of convergence in an ill-conditioned network (large condition number) is extremely slow as the performance function is valley shaped and the weights are slow to decay along the valley floor.

The rate of convergence of steepest descent can also be shown to be related to the condition of the autocorrelation matrix. Therefore, this is a fundamental concept which determines the rate of parameter convergence as well as its generalisation abilities, and is discussed further in the next section.

Network conditioning

The condition of a linear in its parameter vector network (such as an AMN) is determined by the shape of the basis functions as well as the distribution of the training data which in turn produces the elements of the autocorrelation matrix, see Eq.(3.7.11). The amount of overlap in the basis functions and their degree of activation determines the form of the autocorrelation matrix and both of these can be influenced by the designer to produce better conditioned systems. AMN are generally well-conditioned because the basis functions have a unimodal localised shape and so many of these are mutually orthogonal (R is sparse) and the power a basis function ($E(\phi_i^2)$) is generally greater than its interaction with the other basis functions (R tends to be diagonally dominant). So by simple matrix/eigenvalue results based on Gersgorin's circles, AMNs tend to be well-conditioned.

In spite of these arguments, AMNs can become badly conditioned when the order of the basis functions becomes too high (thus increasing the amount of overlap and its support) or the dimension of the input space increases too much or the training data lies at the edge of a basis function's support. All of these considerations must be taken into account when a network is designed and ways of designing the network (automatically) in order to simplify its structure are described in section 4.3.2.

It is interesting to see how these arguments can be applied to other ANN architectures as well. For instance, the MLP can be considered as a particular type of projection pursuit network, except that the projection pursuit algorithm requires that the power $(E(\phi_i^2)))$ of each hidden layer node is equal, which is also equivalent to preconditioning the autocorrelation matrix such that it has equal diagonal elements. In addition, when MLPs' node transfer functions are chosen to be *orthogonal* to the bias term (by using the tanh(.) function rather than the standard $(0, 1)$ sigmoid), the condition of the autocorrelation matrix is improved. So by ensuring that the power of every node in an MLP is equal (including the bias node), and by choosing appropriate sigmoidal-type transfer functions, the condition of the MLP, and hence its rate of parameter convergence, can be improved [19].

Network singularity

Very often in AMNs, the overall system is *singular*[4] as this occurs when a basis function does not have any training samples lying in its support. It can be a serious problem when trying to invert the autocorrelation matrix directly, but although it affects the gradient-based iterative methods it doesn't change their overall behaviour. This can be easily seen by considering the form of a singular performance surface (as shown in Fig. 3.11). A zero eigenvalue corresponds to a flat valley floor and the weights are not updated along this direction. There are an infinite number of global minima and the gradient based methods converge to an optimal solution nearby.

For instance, the dynamical behaviour of v_i which is the projection of the weight vector error onto the ith principal axis which has a zero eigenvalue is given by:

$$v_i(k + 1) = v_i(1) \tag{3.7.36}$$

i.e. it is *not* affected by the training process. Also, as there exist an infinite number of optimal solutions, one could be chosen such that the initial errors, which correspond to the zero eigenvalues, are all zero and the gradient-based algorithms will always converge to this solution. Singularity does not affect the rate of parameter convergence (and this is why $\lambda_{\min}$ is defined as being the smallest, positive eigenvalue in the definition of the network's condition number), it only influences which optimal solution is found.

When R is singular and there exists an infinite number of global solutions, the optimal solution is usually considered to be the *smallest* one as this can be considered to be a form of *regularisation*. This *minimal (Euclidean) norm solution* corresponds to the answer obtained from using the generalised inverse in section 3.7.1. It can be easily shown that whenever the initial weight vector is zero or lies in the space spanned by the transformed

---

[4]The autocorrelation matrix has at least one zero eigenvalue.

input vectors, gradient-based methods will always find the minimum norm solution and this is why many weight vectors are initialised with small (or ideally zero) values.

3.8   Instantaneous Learning Rules

In order for a system to organise in real time, on-line instantaneous training is required and an instantaneous estimate of the network's current performance needs to be evaluated prior to weight updating. The instantaneous MSE, $\epsilon_y^2(k)$, is widely used as it is simple to calculate and only uses the most recent piece of training information. Its gradient is also the (unbiased) instantaneous estimate of the true gradient at time k:

$$- 2\epsilon_y(k)\phi(k) \tag{3.8.1}$$

3.8.1   Least mean squares

Updating the weight vector in proportion to the instantaneous gradient estimate gives the instantaneous gradient descent or Least Mean Squares (LMS) training rule

$$\Delta\theta(k) = \delta\epsilon_y(k)\phi(k) \tag{3.8.2}$$

where δ is a learning rate. This algorithm is extremely simple to implement as a weight is updated by an amount proportional to the current error multiplied by the size of its contribution to the output. When $\phi(k)$ is sparse, as occurs in many AMNs, only those weights that influenced the output are updated and the vast majority are not changed.

After adaption using the LMS rule the *a posteriori* AMN output $\hat{y}(k)$ is given by:

$$
\begin{aligned}
\hat{y}(k) &= \phi^T(k)\theta(k) \tag{3.8.3} \\
&= \delta\|\phi(k)\|_2^2 y(k) + (1 - \delta\|\phi(k)\|_2^2)\hat{y}(k) \tag{3.8.4}
\end{aligned}
$$

and *a posteriori* output error is:

$$
\begin{aligned}
\epsilon_y(k) &= y(k) - \hat{y}(k) \tag{3.8.5} \\
&= (1 - \delta\|\phi(k)\|_2^2)\epsilon_y(k) \tag{3.8.6}
\end{aligned}
$$

where $\|.\|_2$ is the standard Euclidean norm. For a nonzero $\epsilon_y(k)$ the following relationships between the *a priori* and the posteriori output errors can be established for various ranges in the learning rate δ.

$$|\underline{\epsilon}_y(k)| > |\epsilon_y(k)| \qquad \text{if } \delta \notin \left[0, \frac{2}{\|\phi(k)\|_2^2}\right] \tag{3.8.7}$$

$$|\underline{\epsilon}_y(k)| = |\epsilon_y(k)| \qquad \text{if } \delta = 0 \text{ or } \delta = \frac{2}{\|\phi(k)\|_2^2} \tag{3.8.8}$$

$$|\underline{\epsilon}_y(k)| < |\epsilon_y(k)| \qquad \text{if } \delta \in \left(0, \frac{2}{\|\phi(k)\|_2^2}\right) \tag{3.8.9}$$

Clearly the performance of the LMS rule depends directly on the magnitude of the transformed input vector as it influences the stability of the algorithm. For stable learning, δ must satisfy:

$$0 < \delta < \frac{2}{\|\phi(k)\|_2^2} \tag{3.8.10}$$

and this can be easily obtained by noting that $\|\phi(k)\|_2^2$ is the maximum (only non-zero) eigenvalue of the instantaneous performance function. Generally a large δ leads to fast initial learning, whereas a small learning rate ensures greater insensitivity to model mismatch and measurement noise.

3.8.2 Normalised least mean squares

The Normalised Least Mean Squares (NLMS) training algorithm is an instantaneous version of the steepest descent rule and can be obtained through setting:

$$\delta(k) = \frac{\delta'}{\|\phi(k)\|_2^2} \tag{3.8.11}$$

so that

$$\Delta\theta(k) = \frac{\delta'\epsilon_y(k)\phi(k)}{\|\phi(k)\|_2^2} \tag{3.8.12}$$

which is also the NLMS *error correction* algorithm. For $\delta' = 2$, the *a posteriori* output error is always zero (see Eq.(3.8.8)), storing the desired value exactly. The weight change resulting from the NLMS rule is in the same direction as that resulting from the LMS rule, it is only the distance travelled (step size) parallel to the transformed input vector that changes.

When the NLMS learning rule is utilised, output error reduction is independent of the transformed input vector $\phi(k)$, although the dependence of the learning rate, δ, on $\|\phi(k)\|_2^2$ ensures that the NLMS rule no longer minimises the MSE, rather it minimises

$$E\left(\frac{\epsilon_y^2(k)}{\|\phi(k)\|_2^2}\right) \tag{3.8.13}$$

If there exists a unique weight vector such that $\epsilon_y(k) = 0 \ \forall k$ (no modelling error) or if $\|\phi(k)\|_2^2$ is constant for all k, the optimal weight vectors which minimise the MSE and Eq.(3.8.13) are equivalent. Otherwise the minima will occur at different locations in the weight space.

A full theoretical comparison of the LMS and NLMS learning rules and the gradient and steepest descent training algorithms is quite difficult, although some insights can be gained by calculating the condition of the autocorrelation matrix. In section 3.11 it is shown that in certain cases the amount of *gradient noise*[5] is directly proportional to the

---

[5] Gradient noise is defined as being the difference between the true and estimated (instantaneous) gradient.

autocorrelation matrix, so the LMS algorithm can be compared with gradient descent. Similarly, the rate of convergence of the NLMS rule is determined by the *normalised* autocorrelation matrix (the elements are normalised with respect to the strength of the input signal), so the condition of the autocorrelation and its normalised counterpart can be compared to determine the relative rates of convergence of these instantaneous learning rules.

3.8.3   Stochastic approximation

In the LMS and the NLMS training rules, the learning rate δ is assumed to be a positive, constant number. However, when there exists modelling error or measurement noise, these algorithms will only converge if the learning rate is scheduled against time such that it gradually filters out this noise. Stochastic approximation theory provides one method for performing this operation. Let an individual learning rate $\delta_i(k)$ be assigned to each basis function, let it also be a function of time k, then fast initial convergence together with long term noise filtering can be achieved in the so called stochastic approximation versions of the LMS and NLMS algorithms. The necessary conditions on $\delta_i(k)$ for the weight vector to converge are given by:

$$\delta_i(k) \;>\; 0 \qquad\qquad (3.8.14)$$

$$\lim_{k \to \infty} \delta_i(k) \;=\; 0 \qquad\qquad (3.8.15)$$

$$\sum_{k=1}^{\infty} \delta_i(k) \;=\; \infty \qquad\qquad (3.8.16)$$

$$\sum_{k=1}^{\infty} \delta_i^2(k) \;<\; \infty \qquad\qquad (3.8.17)$$

One such function which satisfies these constraints is:

$$\delta_i(k) = \frac{\delta_1}{1 + k_i/\delta_2} \qquad\qquad (3.8.18)$$

where δ_1 and δ_2 are positive constants which denote the initial learning rate and the rate of decay respectively, and k_i is the number of times that the ith basis function has been non-zero.

3.9   Weight Convergence

Like the gradient-based methods, the LMS and NLMS training algorithms can be analysed by looking at the dynamical behaviour of the weight vector error, although there is one important difference: the network only converges when it can reproduce the training data exactly. When there exists modelling error or measurement noise, the network's weights no longer converge to a point but rather to a domain as the unmodelled mismatch provides a persistently exciting term. So the convergence of the weight vector is analysed initially by assuming that there exists a unique solution to the training data. The assumption is unreasonable in practice, but the proof is instructive as it provides insights into the relative rates of convergence of these techniques.

Suppose that there exists a set of (transformed) training examples $\{\phi(k), y(k)\}_{k=1}^{T}$, which are cyclically presented to an AMN trained using the NLMS rule. Also assume that there exists a unique weight vector $\widehat{\theta}$ such that

$$y(k) = \phi^{T}(k)\widehat{\theta} \ \forall k \tag{3.9.1}$$

and the transformed input vectors possess the property that for some positive constant a_{ϵ}

$$0 < a_{\epsilon} \leq \|\phi(k)\|_{2}^{2} \leq 1 \tag{3.9.2}$$

which is generally true for most AMNs. This is in fact the matching system to be described in chapter 5.

Expanding $\Delta\theta(k)$ in Eq.(3.8.12) and subtracting $\widehat{\theta}$ from each side gives:

$$\epsilon_{\theta}(k+1) = \epsilon_{\theta}(k) - \frac{\delta\epsilon_{y}(k)}{\|\phi(k)\|_{2}^{2}}\phi(k) \tag{3.9.3}$$

where $\epsilon_{\theta}(k) = \widehat{\theta} - \theta(k)$, and noticing that $\epsilon_{y}(k) = \phi^{T}(k)\epsilon_{\theta}(k)$ produces:

$$\epsilon_{\theta}(k+1) = \left(I - \delta\frac{\phi(k)\phi^{T}(k)}{\|\phi(k)\|_{2}^{2}}\right)\epsilon_{\theta}(k) \tag{3.9.4}$$

Taking the 2-norm of this expression gives:

$$\|\epsilon_{\theta}(k+1)\|^{2} = \|P(k)\,\epsilon_{\theta}(k)\|^{2} \leq \|P(k)\|^{2}\,\|\epsilon_{\theta}(k)\|^{2} \tag{3.9.5}$$

where $P(k) = (I - \delta\phi(k)\phi^{T}(k)/\|\phi(k)\|_{2}^{2})$ is a *projection operator* with $\delta = 1$. The 2-norm of $P(k)$ is just its maximum eigenvalue, $\lambda_{\max}$, as the matrix is symmetric. Now it can easily be seen that $\phi(k)$ is an eigenvector of $P(k)$ with a corresponding eigenvalue $(1-\delta)$, and every remaining eigenvector has a unity eigenvalue. Therefore:

$$\|P(k)\|^{2} = 1 \tag{3.9.6}$$

and the sequence $\{\|\epsilon_{\theta}(k)\|^{2}\}_{k}$ is non-increasing. At the kth iteration, the component of $\epsilon_{\theta}(k)$ which is parallel to $\phi(k)$ is reduced by an amount $(1-\delta)$ and the remaining orthogonal components (the projection of $\epsilon_{\theta}(k)$ onto the remaining eigenvectors) are the same. Parameter convergence can then be proven provided that the learning rate lies in the interval $(0, 2)$ and that the training samples span the whole weight space (i.e. are persistently exciting).

A similar analysis can be performed for the LMS rule to show that it also converges to the optimal solution, and this is also true for the stochastic approximation versions of these learning algorithms.

Minimal norm solutions

It has been shown that if there exists a unique solution to the set of linear equations which are generated by the AMN training data, the parameter vector will converge to this value when an LMS/NLMS training rule is used. However, when it is assumed that the training data is *consistent*, but not necessarily spanning, there are an infinite number of optimal solutions, but like the gradient based batch training algorithms, the LMS algorithms converge to the minimum norm solution whenever the initial weight vector is zero or else lies in the space spanned by the transformed training inputs. This is because the batch and instantaneous learning rules search the same region of weight space and the minimum norm solution is the only optimal minima that lies in this region, hence the weight vector always converges to this value.

Similarly, when the stochastic LMS algorithms are used, they always find the minimum norm solution even when $J_{\min}$ is non-zero.

3.10   A Geometric Interpretation of the LMS Algorithm

The various LMS learning rules are of the basic form:

$$\Delta\theta(k) = c(k)\phi(k) \tag{3.10.1}$$

and it is only the (step size) scalar term $c(k)$ that determines the rule; in all cases the weight change (search direction) is parallel to the current transformed input vector $\phi(k)$. The aim of the weight update is to ensure that:

$$y(k) = \phi^T(k)\theta(k+1) \tag{3.10.2}$$

This specifies one equation in h variables (the set of linear weights) and so the solutions to this equation lie on the $(h-1)$-dimensional hyperplane, $h(k)$, (in weight space) which is specified by:

$$h(k) = \{\theta \in \Re^h : 0 = \phi^T(k)\theta - y(k)\} \tag{3.10.3}$$

The normal to the solution hyperplane is the vector $\phi(k)$, and the unit normal is given by $\phi(k)/\|\phi(k)\|_2$. In a 2-dimensional weight space, the solutions of such an equation lie on a line (one equation with two variables) as illustrated in Fig. 3.13. The LMS rules update the weight vector parallel to the transformed input vector and the transformed input vector is perpendicular to the solution hyperplane, which means that the weight change is also perpendicular to the solution hyperplane as shown in Fig. 3.13.

The effect of different values of δ being used in NLMS adaption rule can be illustrated geometrically. When $\delta = 1$, the weight update is such that the new weight vector must lie on the solution hyperplane. If $0 < \delta < 1$, then the weight update is such that it

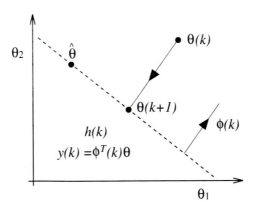

Figure 3.13: Solution hyperplane (dashed line) and perpendicular learning in a 2-dimensional weight space.

does not reach the solution hyperplane (overdamped), whereas if $1 < \delta < 2$ then the weight update vector crosses the solution hyperplane and lies on the opposite side of the hyperplane compared to $\theta(k)$ (underdamped). If $\delta < 0$ then the weight update moves perpendicularly away from the solution hyperplane and if $\delta > 2$ then the weight update moves away from the solution hyperplane on the opposite side (unstable learning). This is illustrated in Fig. 3.14

3.10.1 Principle of minimal disturbance

The LMS family of learning rules embody the principle of *minimal disturbance*, that is to say the weight change made to the weight vector is the smallest one which will cause the new desired output to be stored. Since the weight change is the smallest one of all the possible weight changes which will cause the new weight vector to store the new desired output, it interferes minimally with the stored information.

The minimal distance (with respect to the Euclidean norm), $d(k)$, of the solution hyperplane $h(k)$, from the point, $\theta(k)$ is:

$$d(k) = \frac{|\epsilon_y(k)|}{\|\phi(k)\|_2} \tag{3.10.4}$$

which is achieved exactly by the NLMS rule with $\delta = 1$, i.e. $\|\Delta\theta(k)\| = d(k)$.

The NLMS learning rule can also be derived by viewing it as a constrained optimisation problem, where the constraint ensures that the new weight vector lies on the solution hyperplane and the weight change is minimal with respect to the Euclidean norm, i.e.

$$\text{minimise} \quad \|\Delta\theta(k)\|^2 \tag{3.10.5}$$
$$\text{subject to} \quad y(k) - \phi^T(k)\theta(k+1) = 0 \tag{3.10.6}$$

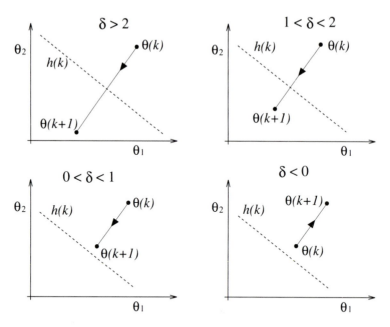

Figure 3.14: The effect on the weight updates for different values of δ in the NLMS adaption rule.

Therefore the minimal disturbance principle is automatically incorporated into the learning law. This optimisation problem can be solved using Lagrange multipliers, and (not surprisingly) the recommended change in the weight vector is simply parallel to the transformed input vector. Similarly, the NLMS learning rule can be formed as an SVD problem which automatically incorporates the minimal disturbance principle.

Recently, a set of modified NLMS-type learning rules have been proposed [44] that are derived from a generalised set of minimal disturbance conditions:

$$\text{minimise} \quad \|\Delta\theta(k)\|\, p \tag{3.10.7}$$
$$\text{subject to} \quad y(k) - \phi^T(k)\theta(k+1) = 0 \tag{3.10.8}$$

where $p \in [1, \infty]$. However, the resulting training algorithms can become unstable, especially for $p = 1$ or ∞ [18], and the NLMS rule which is generated with $p = 2$ is both remarkably simple and robust.

3.10.2 Weight convergence related to basis structural dependency

The principle of minimal disturbances ensures that the new weight vector is *close* to the previous weight vector after adaptation, but it does not necessarily imply that important information won't be "forgotten" or overwritten. This topic is now addressed by investigating how the rate of parameter convergence depends on the direction of the weight updates and its implications for the design of the nonlinear, topology conserving map $\xi \to \phi$.

Consider the case when two successive transformed input training vectors $\phi(1)$, and $\phi(2)$ are mutually orthogonal, then the output error for the first pattern after updating the second is:

$$\epsilon_y(3) = \phi^T(1) \left(w(2) + \frac{\delta\epsilon_y(2)\phi(2)}{\|\phi(2)\|_2^2} \right) \qquad (3.10.9)$$

but $\phi^T(1)\phi(2) = 0$, hence the output error for the first pattern is not changed when information about an orthogonal datum is stored. The weights are updated in orthogonal directions, and indeed, when the whole training set consists of orthogonal transformed input vectors, parameter convergence occurs in a finite time with $\delta = 1$. When the training patterns are orthogonal, training patterns do not interfere with each other.

The fast initial learning occurs in AMN because the transformed input vectors are *sparse*. Hence many of the transformed input vectors are mutually orthogonal and so learning about one part of the input space will not interfere with learning about another part of it; this is a major attribute for *online* modelling and control.

If instead the transformed input vectors are nearly parallel (*i.e.* are ill-conditioned) then it is not difficult to demonstrate that the weight error reduction changes little from iteration to iteration, see Fig. 3.15. This might suggest that it is always desirable to have the transformed input vectors $\phi(k)$ in an AMN orthogonal to each other. This can only be achieved if the weight set mapped to by each input is unique, or there exists at most one input lying in the receptive field of each basis function. However, to locally generalise, each basis function needs an input lying in its receptive field. If the training samples are corrupted by noise it may be necessary for several samples to lie in the receptive field of each basis function. Finally, having very small receptive field widths would lead to enormous memory requirements. The size and shape of the basis functions is a set of design parameters which influence the modelling capabilities, the computational cost and the rate of convergence.

It is important to study parameter convergence rather that simply measuring the instantaneous output errors as this provides a measure of the systems ability to generalise. Consider the case when a discrete time single input/output first order recurrent system is subject to a sinusoidal input signal. The training data is generated by sampling the plant input/output data so that the solution hyperplane are at angles of $10°$ relative to each other (*i.e.* 18 samples in total) as shown in Fig. 3.15, where the weight updates are generated by cyclic training (solid line) and random training (dotted line). When the training data is sequentially presented to the network, parameter error decays by $\cos^2(10°)$ at iteration, parametric convergence is slow as successive training examples are ill-conditioned. By randomly choosing the training pair from the data set, parameter convergence occurs quicker because successive training pairs are more orthogonal. This

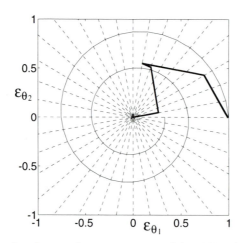

Figure 3.15: Solution hyperplanes are rotated by $10°$ relative to each other.

implies that using instantaneous learning rules for on-line plant modelling and control may produce systems which have poor generalisation abilities unless special care is taken to ensure that the distribution of the training signals are appropriate (persistently exciting).

3.11   Gradient Noise

Online weight adjustment rules have been developed on the basis of instantaneous gradient-based methods, from which it was shown that the rate of parametric convergence depends on the condition number, $C(R)$, of the correlation between transformed input samples. As instantaneous LMS (or NLMS) is typically used for online modelling and control, an error, $n(k)$, is introduced by using the instantaneous rather than the batch performance gradient, such that:

$$n(k) = \nabla(k) - \widehat{\nabla}(k) \qquad (3.11.1)$$

where $\widehat{\nabla}(k)$ is the instantaneous approximation to the true gradient $\nabla(k)$ at time k. This difference in the gradients is also called *gradient noise*, and its covariance matrix describes the relative spread of its components. When the input samples are Gaussian distributed with zero-mean and independent [4], it can be shown that:

$$N(k) = R\epsilon_\theta \epsilon_\theta^T R + R(J_{\min} + \epsilon_\theta^T R\epsilon_\theta) \qquad (3.11.2)$$

where $N(k)$ is the gradient noise covariance matrix which is strongly dependent on the form of the autocorrelation matrix. This is also true when the inputs are non-zero mean.

When θ is near its optimal value $\widehat{\theta}$, then $N \simeq J_{\min} R$. The above simplified expression for gradient noise covariance is influenced by the gradient correlation matrix, which causes the orientation if the principal axes defined by N and R to be different (except when

$\theta \rightarrow \hat{\theta}$). However, when $\epsilon_\theta(k)$ lies on one of the principal axes of R, N can be diagonalised using the orthonormal matrix of R, resulting in the strongest gradient noise component along the steepest principal axis of R. Thus the gradient noise weight fluctuation is maximum along the flattest axis, and minimum along the axis with smallest curvature. Fig. 3.16 illustrates the more usual case for two weight convergent sequences using the LMS algorithm when initial weights error vectors lie on different principal axes.

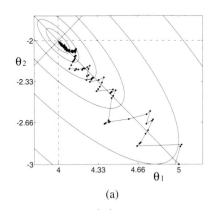 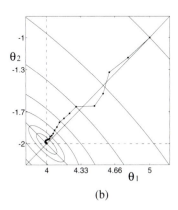

(a)   (b)

Figure 3.16: Typical weight convergence properties of the LMS adaption. (a) along the flattest axis; (b) along the steepest axis.

3.11.1 Minimal capture zones

When the training data cannot be interpolated exactly by a network, the gradient noise provides a non-vanishing excitation term in the weights update equation close to their optimal values. This causes the weights to converge to a *domain*, rather than to a unique point, whose size and shape depends on the type and size of the modelling error as well as the internal representation used by the AMN.

Parameter convergence to a domain rather than to a point occurs because an instantaneous network output error may be due to either parameter errors or optimal modelling error/measurement noise. This can be seen by writing the instantaneous output error as:

$$\epsilon(k) = \phi^T(k)\left(\hat{\theta} - \theta(k)\right) + \left(y(k) - \phi^T(k)\theta(k)\right) \qquad (3.11.3)$$
$$= \epsilon^o(k) + \epsilon^n(k) \qquad (3.11.4)$$

where $\epsilon^o(k)$ is the difference between the current and the optimal network and $\epsilon^n(k)$ is the optimal modelling error or measurement noise. When an instantaneous learning rule is used to adapt the coefficient vector it is generally unable to distinguish between these two quantities, therefore a considerable amount of noise is injected into the dynamical behaviour of the weight vector, especially close to its optimal solution.

As an example, consider a linear network which has a single adjustable parameter and an output given by $\hat{y}(k) = \phi(k)\theta(k)$. The training set is composed of two input/output pairs given by $\{\phi(1), y(1)\} = \{1, 1\}$ and $\{\phi(2), y(2)\} = \{2, 4\}$, and the associated instantaneous and true cost functions are shown in Fig. 3.17. If $\theta(k) < 1$ or $\theta(k) > 2$, both

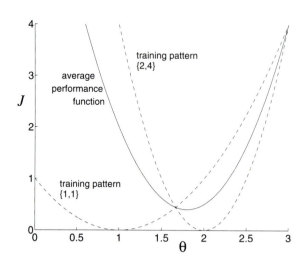

Figure 3.17: The true (solid line) and the instantaneous MSE performance functions (dashed lines) which correspond to the input/output data $\{1, 1\}$, $\{2, 4\}$.

instantaneous gradient estimates are consistent (point in the same direction). However, when $\theta(k) \in (1, 2)$ the gradient estimates formed from the instantaneous data are inconsistent and the parameter $\theta(k)$ moves about in this interval and does not converge to a unique value. When there is more than one adjustable parameter the weight vector converges to a p-dimensional domain as will now be described.

This minimal capture zone for a set of data randomly drawn from a possibly infinite training set where an NLMS learning rule is used with $\delta \in (0, 1)$, can be defined as:

> *The smallest n-dimensional domain which contains the intersection of all the solution hyperplanes such that the orthogonal projection of every point (in the domain) onto every solution hyperplane is completely contained in this domain.*

A 2-dimensional minimal capture zone is shown in Fig. 3.18. The size and shape of these minimal capture zones depends on the training samples, the type of modelling error, the order of presentation, the adaptation rule and the learning rate δ. Without discussing all of these aspects in great detail it is worthwhile noting that the minimal capture zones which correspond to random measurement noise can be *infinite* whereas the ones that are generated by modelling error are generally *finite*. The smaller the learning rate the greater is the noise filtering (equivalently the modelling error filtering), although the

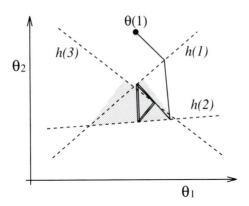

Figure 3.18: A 2-dimensional minimal capture zone generated by three inconsistent training samples.

capture zone is not always smaller. When the learning rate is chosen so as to comply with the stochastic approximation conditions and an LMS updating rule is used, convergence of the weight vector to the optimal MSE value is assured (and similarly for the stochastic NLMS learning rule which minimises the normalised MSE).

3.11.2 Output deadzones and leaky integrators

Two ways of overcoming the *parameter leakage* which is due to either modelling error or measurement noise is to introduce output deadzones and leaky integrators into the instantaneous weight update equations. Both techniques result in biased solutions but they aim to produce acceptable answers in a more robust manner.

Output deadzones attempt to subtract the optimal modelling error from the instantaneous output error in the following manner:

$$\epsilon_y^d(k) := \left\{ \begin{array}{ll} 0 & \text{if } |\epsilon_y(k)| \leq \zeta \\ \epsilon_y(k) + \zeta & \text{if } \epsilon_y(k) < -\zeta \\ \epsilon_y(k) - \zeta & \text{if } \epsilon_y(k) > \zeta \end{array} \right. \tag{3.11.5}$$

where $\epsilon_y^d(.)$ is the modified output error which is used in the instantaneous learning algorithms and ζ is an estimate of the maximum (or averaged) $\epsilon^n(k)$. This modified output error introduces a ball of radius ζ around the optimal parameters' value and after the weights enter this region they are not modified any further. However, this can cause premature parameter convergence, especially when $\epsilon^n(k)$ has been inaccurately estimated.

Leaky integrator learning rules attempt to counteract parameter drift by adding a term of the form $-\delta_2 w$ onto the right hand side of the LMS weight update equations. This term prohibits the weights from becoming too large and introduces a form of *regularisation* into the adaptive model. As with the output deadzones, the choice of the learning rate δ_2 can significantly affect the performance of this algorithm and it could also be argued that in order to retain some temporal stability when ϕ is sparse, the ith component of the leaky integrator should be formulated as $-\delta_2\theta_i\phi_i$.

Chapter 4

Fuzzy Modelling and Control Systems

Fuzzy logic was introduced by L. Zadeh in the mid-sixties as a method for representing and utilising vague or uncertain knowledge [176, 177]. He reasoned that a new approach was necessary to humanise the increasingly complex systems being developed, and stated this in his *principle of incompatibility*:

> *As the complexity of a system increases, our ability to make precise and yet significant statements about its behaviour diminishes until a threshold is reached beyond which precision and significance (or relevance) become almost mutually exclusive characteristics.*

Fuzzy logic deals with the manipulation and utilisation of vague, natural language statements and expressions. It allows common-sense, heuristic rules to be encoded within a fuzzy system and also means that a learning module can explain its actions using a set of fuzzy rules. Fuzzy systems are therefore *transparent*[1] to the user and this is in direct contrast to many Artificial Neural Networks (ANNs).

Like many Artificial Intelligence (AI) algorithms (indeed fuzzy logic can be considered as an AI technique), fuzzy logic was originally applied to those tasks which humen perform apparently effortlessly but are difficult for conventional algorithmic based methods: speech recognition, partial pattern matching, visual data analysis, etc. The development of fuzzy logic was inspired by the way humen make decisions using incomplete, inaccurate information, [87], but it is in the area of systems engineering where most of the fuzzy products have appeared [46]. This is probably because most fuzzy modelling and control systems use only a small number of well-understood fuzzy information processing techniques, hence it is possible to gain informative insights into the learning and generalisation abilities of these algorithms.

---

[1]Knowledge is stored in a form which can be easily understood by the designer.

The use of fuzzy logic in control engineering has had a chequered history since it was originally applied by A. Mamdani and his students at Queen Mary College, twenty years ago [106]. During the seventies, a number of static fuzzy controllers were developed and the first self-organising fuzzy controller was produced which was able to change its rules in order to improve its performance [130]. Research died off in the United Kingdom during the early eighties, but in Japan and the Far East a significant number of major electronic and automotive companies became interested in the subject and during the latter part of the decade, and the early nineties, a large number of consumer products appeared (washing machine and temperature regulation controllers, engine management systems, etc.) which are based on fuzzy logic. The majority of these systems are *static* and the success of these fuzzy systems is due to the way fuzzy logic represents and manipulates vague, expert knowledge in order to produce complex nonlinear control surfaces. An iterative design procedure can be used to develop these systems, after being initialised with a sensible set of rules; but it important to realise that the majority of fuzzy systems currently being developed and applied are static. The ability of a static fuzzy system to resolve the ambiguity associated with heuristic rules is its most important property.

Fuzzy systems process "seemingly" vague or imprecise expressions such as *the input is small* when they are used to draw useful conclusions from inexact data. However, the word seemingly has been placed in quotations because, in order to represent a collection of fuzzy statements and rules on a computer, any ambiguity or vagueness associated with the original statement must be completely resolved. It may be hidden from the end-user or partially known by the expert who initialises the system, but any vagueness must be given an exact domain-specific interpretation. For instance, consider the following statement:

the bike is moving fast

It provides relative information about the speed of the bike but in order for this knowledge to be useful, domain-specific facts (such as the type of bike) must be known. The statement has different meanings depending on whether its a pedal cycle used for shopping, a moped or a 500cc racing motorbike and there are other factors which could influence its definition i.e. the weather (foggy, raining or dry). Often this type of domain knowledge is assumed to be implicitly known, but it must be explicitly given to a fuzzy system, either through the definition of the set *fast*, or by including an extra variable which specifies the type of bike.

Fuzzy logic, and the fuzzy systems which use it, is an extremely useful tool for many difficult modelling and control systems, and this chapter highlights these points. It also illustrates when there may be problems with applying these techniques (such as the curse of dimensionality) and potential solutions for overcoming these difficulties.

4.1 Fuzzy Systems

A fuzzy system uses fuzzy logic in any of its initialisation, validation or reasoning processes. An expert can initialise a fuzzy system by writing down a set of vague production rules known as a fuzzy algorithm, similarly a trained fuzzy system may explain its actions using a fuzzy algorithm. When a fuzzy system uses fuzzy logic in its reasoning process, any of the inherent imprecision associated with this vague representation is completely resolved and its input/output behaviour is well defined. This section investigates how a fuzzy system may be implemented and shows that when the fuzzy logic uses algebraic (sum/product) operators and the fuzzy sets are represented using B-spline or Gaussian basis functions, the overall inferencing operations are considerably simplified and the fuzzy system is equivalent to a B-spline or Gaussian Radial Basis Function (GRBF) neurofuzzy network[2] discussed in chapter 3. The overall neurofuzzy system simply operates as a nonlinear numerical processing device but it has the advantage that it can be initialised and interpreted using a set of vague linguistic rules.

4.1.1 Fuzzy algorithms

A fuzzy *algorithm* is composed of a collection of vague statements of the form:

IF (ξ_1 *is positive small* AND ξ_2 *is negative medium*)
THEN (y *is negative small*) (0.5)
OR IF (ξ_1 *is almost zero* AND ξ_2 *is postive large*)
THEN (y *is positive small*) (1.0)
OR IF (ξ_1 *is negative large* AND ξ_2 *is postive small*)
THEN (y *is negative medium*) (0.7)

Each fuzzy production rule relates a set of vague linguistic terms which represent the inputs to the algorithm to another set that refer to the algorithm's output. The number in brackets that follows each rule represents the confidence in that rule being true, where a value of unity denotes a completely true rule and a value of zero means that the rule never contributes to the algorithm's the output. Values lying in between represent varying degrees of truthfulness about the statement.

A fuzzy algorithm represents a human's *subjective* and *context dependent* interpretation of a particular input/output mapping and does not indicate how the vague linguistic terms, such as *smallness*, should be represented. Similarly, it does not specify how the fuzzy operators (intersection, union, implication), should be implemented in a physical system. Once a specific definition has been given to these vague or fuzzy concepts, thus

---

[2]A device that combines the architecture and learning properties of a neural network with the representational advantages of a fuzzy system.

resolving any implicit subjectivity, the overall input/output mapping is completely deterministic as there is no longer any uncertainty associated with their definitions. Fuzzy logic deals with the manipulation and representation of these vague terms.

The ability to represent a nonlinear mapping as a set of linguistic rules is arguably the most important feature of a fuzzy system. It means that a fuzzy system can be initialised using an expert's heuristic knowledge which is expressed using natural language terms. It also makes an adaptive fuzzy system *transparent,* as it provides the designer with significant information about the learnt function.

4.1.2 Systems

A *fuzzy system* contains all the components necessary to implement a fuzzy algorithm and resolve all of the associated vagueness. It is composed of four basic elements:

- **a knowledge base** which contains definitions of the fuzzy sets and the fuzzy operators;

- **an inference engine** which performs all the output calculations;

- **a fuzzifier** which represents the real valued inputs as fuzzy sets; and

- **a defuzzifier** which transforms the fuzzy output set to a real valued output,

and this is illustrated in Fig. 4.1. The knowledge base contains the definitions of each of the fuzzy sets and maintains a store of operators used to implement the underlying logic (AND, OR etc.), as well as a *rule confidence matrix* which represents the fuzzy rule mappings. The inference unit, together with the fuzzifier and the defuzzifier allows real-valued outputs to be calculated from real valued inputs. The fuzzifier represents the input as a fuzzy set which allows the inferencing unit to match it against the antecedents of the rules stored in the knowledge base. Then the inferencing unit calculates how strongly each rule fires and outputs a fuzzy distribution (union of all the fuzzy output sets) that represents its fuzzy estimate of the true output. Finally, this information is defuzzified (compressed) into a single value which is the output of the fuzzy system.

These systems are extremely flexible and can be used as a basic plant model, a controller, an estimator, or to represent a performance function or as a desired trajectory generator. They implement a general nonlinear mapping and as such can be used for many approximation or classification tasks depending on how the inputs and outputs are chosen.

4.1.3 Fuzzy sets

To represent a linguistic term such as *smallness,* Zadeh proposed the concept of a fuzzy set. A classical Boolean (binary) set has associated with it a *characteristic function* which

fuzzy system

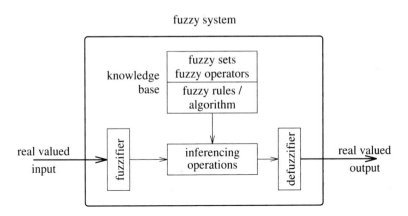

Figure 4.1: A fuzzy system is composed of a knowledge base, an inference engine, a fuzzifier and a defuzzifier.

is one when the element is a member of the set and zero otherwise. Thus the set can be described either explicitly by writing down its members or implicitly by specifying its characteristic function. Zadeh extended the idea of a binary characteristic function such that it allows elements to be a partial member of a set and the *membership function* returns a value which lies in the unit interval. Thus inputs can be partial members of a set and this soft information representation scheme delays any hard (binary) decisions until later in the inferencing or defuzzification processes.

More formally, a fuzzy set A is defined on an input ξ and is characterised by its membership function $\mu_A(.) : \xi \to [0, 1]$. The input domain is either discrete or continuous but it has been shown that the continuous representation is more appropriate for many modelling and control applications [19]. Throughout this chapter the terms "fuzzy set" and "membership function" are used interchangeably.

As is shown in Fig. 4.2, there are a lot of seemingly different fuzzy set shapes, and each one is determined by a set of parameters which, loosely speaking, determine the sets' centres and widths. The two most common choices for the membership functions are the B-spline basis functions and Gaussians.

B-spline membership functions

B-spline basis functions (see Eqs. (3.4.10)-(3.4.12)) are simply piecewise polynomial mappings that have been widely used in the surface fitting community for the past twenty years. The basis functions can be used to represent a set of fuzzy membership functions whose shape is controlled by a group of parameters which is termed the *knot vector*. Many fuzzy designers have been implicitly using order 2 B-spline basis functions when they have implemented their linguistic terms using triangular membership functions, and this is illustrated in Fig. 4.3. The set of knots $\{\lambda_i\}$ determines the width and position of

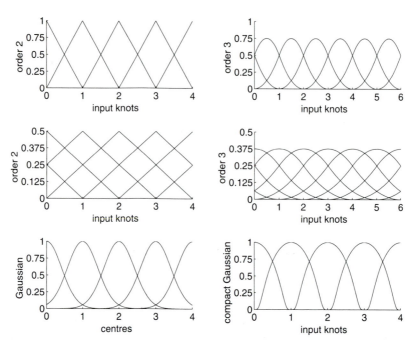

Figure 4.2: Six different possible types of continuous fuzzy sets: B-splines (top), dilated B-splines (middle), Gaussian and compact support Gaussian (bottom).

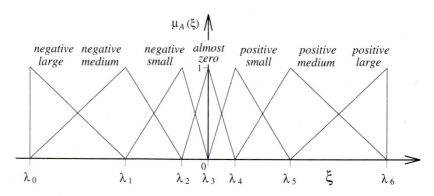

Figure 4.3: A typical set of seven triangular (B-splines of order 2) fuzzy membership functions.

these triangular fuzzy sets (each one is uniquely defined on $k + 1$ successive values) and influences the modelling and learning abilities of the adaptive fuzzy system. It forms a convenient set of parameters for storing the definition of the fuzzy sets and is extremely useful in the output calculations, as a simple algorithm can be used to determine which fuzzy input sets are non-zero for a particular input.

More general order k B-spline basis functions can model *crisp* (non-fuzzy) statements ($k = 1$) and can produce smoother, quadratic and cubic, mappings when order 3 and 4 basis functions are used, respectively, as shown in Figs. 3.8 and 4.2. Also, when the sets are joined together using the addition operator to represent a logical union (OR), the commonly used trapezoidal and Π fuzzy membership functions are formed. The B-spline fuzzy membership functions $\phi_{k,i}(\xi)$ also possess the following properties:

- they are piecewise polynomials of order k;

- they are defined on a compact support as their output is non-zero only in the k intervals;

- they form a partition of unity, ie. $\sum_{i=1}^{h} \phi_{k,i}(\xi) \equiv 1$; and

- their evaluation algorithm is stable and efficient.

The fact that the B-spline fuzzy membership functions are piecewise polynomials means that the network is extremely flexible and certain B-spline based fuzzy systems will be shown to be *universal approximators*[3]. The compact support property means that only a small number of rules fire as only a small number of membership values are non-zero. It also introduces a degree of network stability into the learning algorithm as training about one part of the input space does not significantly affect knowledge stored in a dissimilar region. A set of basis functions which forms a partition of unity is *self-normalising* as it will be shown that the fuzzy system defuzzification process implicitly imposes a partition of unity on the network, and this can significantly affect the form of the fuzzy membership functions. The recurrence relationship that is used to calculate the output of the membership function is given in Eq.(3.4.12), and by choosing the position of the knots appropriately, the basis functions can be designed to vary significantly in areas where the desired function changes rapidly. Also, by specifying multiple knots to occur at the same position, discontinuities in the data can be modelled.

One possible restriction of the normal B-spline fuzzy membership functions is that their order is equal to the size of their support. When wide basis functions are required, this would mean that high-order, overly flexible sets would be used. *Dilated* B-splines decouple this relationship as all that is required is that the size of the support is an integer multiple of the basis function's order, see Fig. 4.2. Like the normal B-spline basis functions, their dilated counterparts all satisfy the four previously mentioned desirable properties.

---

[3]They can approximate any continuous nonlinear system arbitrarily closely on a compact domain.

Gaussian membership functions

Another set of which have been extensively used to represent fuzzy membership functions are the Gaussian functions defined by:

$$\mu_{A^i}(\xi) = \exp\left(-\frac{(c_i - \xi)^2}{2\sigma_i^2}\right) \qquad (4.1.1)$$

These basis functions are easy to implement and the centres (c_i) and widths (σ_i) form a convenient set of parameters to initialise and train. They are also very flexible as by choosing the centres and widths appropriately, the network's output can be made to approximate locally linear functions as well as sigmoidal-type mappings. The Gaussian functions are also defined locally but, strictly speaking, they do not possess a compact support. A modified Gaussian-type membership function that possesses a compact support has been proposed [165] and is defined by:

$$\mu_{A^i}(\xi) = \begin{cases} \exp\left(-\frac{(\lambda_2 - \lambda_1)^2/4}{(\xi - \lambda_1)(\lambda_2 - \xi)}\right) & \text{if } \xi \in (\lambda_1, \lambda_2) \\ 0 & \text{otherwise} \end{cases} \qquad (4.1.2)$$

when the fuzzy set's support is (λ_1, λ_2). This function has a maximum at $\xi = (\lambda_2 + \lambda_1)/2$, where its value is $\exp(-1)$, and is shown in Fig. 4.2.

4.1.4 Operators

In order to implement the fuzzy rules described in section 4.1.1, a set of fuzzy operators which generalise the conventional Boolean intersection (AND), union (OR) and implication (IF THEN) must be selected. Zadeh originally used the min and max operators as they are simple to implement, are equivalent to the Boolean operators for binary arguments and always give results which lie in the unit interval whenever their arguments do. These *truncation* operators were used almost exclusively during the seventies and most of the eighties, and it is only recently that alternative operators have been seriously considered, [19]. Therefore, this section describes *algebraic* operators (sum and product) and provides many reasons for their adoption, whilst in the next chapter algebraic sum/product operators are discussed for the constrction of a learning algorithm for a class of unknown nonlinear systems.

Fuzzy intersection

The antecedent of a fuzzy production rule is formed from the fuzzy intersection (AND) of n univariate linguistic statements:

$$(\xi_1 \ is \ \mu_{A_1^i}) \ \text{AND} \ \cdots \ \text{AND} \ (\xi_n \ is \ \mu_{A_n^i})$$

which produces a new *multivariate* membership function $\mu_{A_1^i \cap \cdots \cap A_n^i}(\xi_1, \ldots, \xi_n)$ or $\mu_{A^i}(\xi)$ defined on the original n-dimensional input space and whose output is given by:

$$\mu_{A^i}(\xi) = \widehat{\prod}\left(\mu_{A_1^i}(\xi_1), \ldots, \mu_{A_n^i}(\xi_n)\right)$$

where $\widehat{\prod}$ is a class of functions called *triangular norms*. Triangular norms provide a wide range of functions to implement intersection of which the two most popular are the min and the product operators. A 2-dimensional fuzzy membership function formed from the product of two triangular (B-splines of order 2) univariate membership functions is shown in Fig. 4.4. Obviously, the shape of the multivariate fuzzy membership function

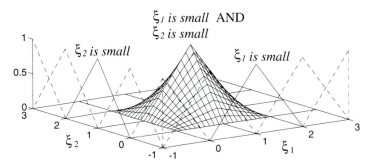

Figure 4.4: A two-dimensional fuzzy membership function.

depends on both the shapes of the univariate membership functions and the operator used to represent the triangular norm. The multivariate membership functions formed using the product operator retain more information than when the min operator is used to implement the fuzzy AND because the latter scheme only retains one piece of information whereas the product operator combines n-pieces. Using the product operator also allows error information to be back propagated through the network as the first derivative is well-defined, it generally gives a smoother output surface (as will be shown later) and when univariate B-spline and Gaussian fuzzy membership functions are used to represent each linguistic statement, the multivariate membership function is simply an n-dimensional B-spline or Gaussian basis function.

Basis function distribution

When all possible fuzzy intersections are taken of n sets of fuzzy membership functions, this implicitly generates an n-dimensional lattice in the original input space on which the multivariate fuzzy membership functions are defined. As illustrated in Fig. 4.5, where a complete set of 2-dimensional fuzzy membership functions are generated by two sets of triangular, univariate fuzzy sets. The bold circles in this figure denote their centres and the shaded area illustrates how two univariate fuzzy sets are combined using the intersection operator. When the fuzzy intersection is taken of every possible combination of univariate fuzzy input sets, the number of multivariate fuzzy membership functions is

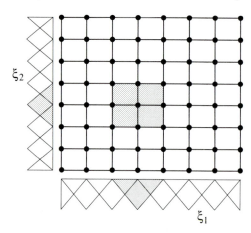

Figure 4.5: A complete set of 2-dimensional fuzzy membership functions.

an *exponential* function of the number of input variables. Such a fuzzy system is said to be *complete*, as for each input there exists at least one multivariate fuzzy set which has a non-zero degree of membership. Removing just one of the multivariate fuzzy sets in Fig. 4.5 would mean that the rule-base was no longer complete as the membership of every basis function is zero at the centre of the missing set. Therefore, unless special techniques are used to structure the inputs to a fuzzy network, these systems suffer from the *curse of dimensionality* which limits their application to small-dimensional modelling and control problems.

Fuzzy implication

To represent a relation (IF *antecedent* THEN *consequent*), let the rule that maps the ith multivariate fuzzy input set A^i to the jth univariate output set B^j with a confidence c_{ij} be labelled by r_{ij}, i.e.:

$$r_{ij}: \qquad \text{IF } (\xi \text{ is } A^i) \text{ THEN } (y \text{ is } B^j) \qquad (c_{ij})$$

Then the degree to which element x is *related* to element y is represented by the $(n+1)$-dimensional membership function $\mu_{r_{ij}}(\xi, y)$ defined in the product space $A_1 \times \cdots \times A_n \times B$ by:

$$\mu_{r_{ij}}(\xi, y) = \hat{*}\mu_{A^i}(\xi)\hat{*}c_{ij}\hat{*}\mu_{B^j}(y) \qquad (4.1.3)$$

where $\hat{*}$ is the binary triangular norm usually chosen to be the min or the product operator. The fuzzy set $\mu_{r_{ij}}(\xi, y)$ represents the confidence in the output being y given that the input is ξ for the ij^{th} fuzzy rule.

For these applications, fuzzy implication can be regarded as an intersection between input and output sets. This is not the only possible interpretation, but is adopted in the vast majority of fuzzy modelling and control systems.

Fuzzy union

If h multivariate fuzzy input sets A^i map to q univariate fuzzy output sets B^j, there are hq overlapping $(n + 1)$-dimensional membership functions, one for each relation. The hq relations may then be connected to form a fuzzy rule base R by taking the union (OR) of the individual membership functions, and this operation is defined by:

$$\mu_R(\xi, y) = \widehat{\sum}_{i,j} \mu_{r_{ij}}(\xi, y) \qquad (4.1.4)$$

where $\widehat{\sum}$ is a class of functions called the *triangular co-norm*. Triangular co-norms also provide a wide range of suitable functions but the two most popular are the max and the addition operators. Using the max operator ensures that effectively only one rule contributes to the output at each particular point in the input/output space whereas summing the individual contributions ensures that several rules influence the *relational surface* $\mu_R(\xi, y)$. However, using the addition operator cannot always be theoretically justified as it may produce outputs which are larger than one. This is not a problem when the input and output univariate fuzzy membership functions form a partition of unity and the product operator is used for intersection and implication and the rule confidence vectors sum to unity, as the system is self-normalising. Similarly, the implicit normalisation performed by the various defuzzification processes can also be used with the addition operator *even* when it produces a value of $\mu_R(\xi, y)$ that is greater than one.

The union of all the individual relational membership functions forms a *ridge* in the input/output space which represents how individual input/output pairs are related and can be used to infer a fuzzy output membership function given a particular input measurement; a process known as fuzzy inferencing. A typical relational surface is shown in Fig. 4.6, where four triangular fuzzy sets (B-splines of order 2) are defined on each variable and the algebraic functions are used to implement the logical operators. This produces a fuzzy relational surface which is piecewise linear between rule centres and the general trend of the input, output relationship is obvious from the contour plot.

4.1.5 Fuzzy rule confidences

The knowledge base in a fuzzy system is comprised of the definitions of the fuzzy membership functions, the fuzzy logical operators and also the fuzzy *rule confidences matrix* C which is of size $(h \times q)$, where h is the number of multivariate fuzzy input sets and q is the number of univariate fuzzy output sets. Each element c_{ij} ($\in [0, 1]$) of the rule confidence matrix represents the strength with which, or confidence in, the ith multivariate fuzzy input set relates to the jth univariate fuzzy output set. When a rule confidence is zero,

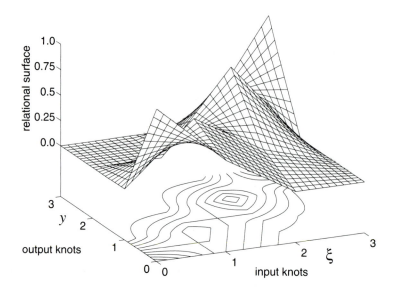

Figure 4.6: A fuzzy relational surface, $\mu_R(\xi, y)$.

it means that the output set never influences the output for that particular fuzzy input set. Alternatively, when a rule confidence is greater than zero, the output set influences the system's output whenever an input partially satisfies the rule's antecedent. Once the fuzzy sets have been defined, the rule confidences encapsulate the expert's knowledge about a particular process and they also form a convenient set of parameters to train.

The rule confidences depend neither on the shape or form of the fuzzy sets nor on the fuzzy logical operators, both of which are stored *separately* in the knowledge base. Discrete fuzzy systems, which have been widely used in self-organising controllers, construct a relational matrix which completely characterises the knowledge base as is implicitly contains information about the fuzzy set shapes, logical operators and rule confidences [65, 130]. Storing knowledge in a distributed fashion as has been described is preferable as it makes it easier understand how differing implementation methods will affect the system's output.

Associated with each multivariate fuzzy input set is a rule confidence vector c_i which represents the estimated output of the system for that particular input set. These rule confidence vectors are generally normalised (sum to unity) as this implies that there is total knowledge about the system's output for that particular input set. These parameters can be easily updated when the knowledge in the rule base is changed. In a lot of adaptive fuzzy systems, the fuzzy output membership functions are altered by shifting their centres which amount to redefining the designer's subjective interpretation of a linguistic statement. It could therefore be argued that these adaptive fuzzy systems cannot be validated after training because the form of the fuzzy sets is not consistent with their original definition. However, when rule confidences are used and stored separately from the fuzzy sets, it is possible to adapt the strength with which a rule fires and still retain its original fuzzy, linguistic interpretation.

4.1.6 Inferencing

Inferencing is the process of reasoning about a particular state, using all available knowledge to produce the best estimate of the output. In a fuzzy system, the inference engine is used to pattern match the current fuzzy input set $\mu_A(\xi)$ with the antecedents of all the fuzzy rules and to combine their responses, producing a single fuzzy output set $\mu_B(y)$. This is defined by:

$$\mu_B(y) = \widehat{\sum}_\xi (\mu_A(\xi) \widehat{*} \mu_R(\xi, y)) \tag{4.1.5}$$

where the triangular co-norm $\widehat{\sum}_\xi$ is taken over *all* possible values of ξ, and the triangular-norm computes a match between two membership functions for a particular value of ξ. When $\widehat{\sum}$ and $\widehat{*}$ are chosen to be the integration (sum) and the product operators, respectively, then:

$$\mu_B(y) = \int_D \mu_A(\xi) \, \mu_R(\xi, y) \, d\xi \tag{4.1.6}$$

which for an arbitrary fuzzy input set requires an n-dimensional integral to be evaluated over the input domain D. The calculated fuzzy output set depends on the fuzzy input set $\mu_A(.)$, the relational surface $\mu_R(.)$ as well as the actual inferencing operators.

As long as there exists an overlap between the fuzzy input set and the antecedents of the rule base, then the fuzzy system is able to *generalise* in some sense. The ability to generalise information about neighbouring states is one of the strengths of fuzzy logic, but their actual interpolation properties are poorly understood. The fuzzy systems studied in this chapter are particularly important as their approximation abilities can be both determined and analysed theoretically which has many important consequences for practical systems.

4.1.7 Fuzzification and defuzzification

The fuzzification and defuzzification processes are the modules which enable the fuzzy rule base to interface with the real-world. A real-valued input must be represented as a fuzzy set in order to perform the inferencing calculations and the information contained in the fuzzy output set must be compressed to a single number which is the real-valued output of the fuzzy system.

Fuzzification

The process of representing a real-valued signal as a fuzzy set is known as fuzzification and is necessary when a fuzzy system deals with real-valued inputs. There are many different methods for implementing a fuzzifier but the most commonly used one is the *singleton* that maps the input ξ^s to a binary or *crisp* univariate fuzzy set A with membership:

$$\mu_A(\xi) = \begin{cases} 1 & \text{if } \xi = \xi^s \\ 0 & \text{otherwise} \end{cases} \tag{4.1.7}$$

For inputs that are corrupted by noise, the shape of the fuzzy set can reflect the uncertainty associated with the measurement process. For example, a triangular fuzzy set may be used where the vertex corresponds to the mean of some measurement data and the base width is a function of the standard deviation. If the model input is a linguistic statement, a fuzzy set must be found that adequately represents the statement. Unless the input is a linguistic statement, there is *no* justification for fuzzifying the input using the same membership functions used to represent the linguistic statements such as *x is small*. The latter membership functions are chosen to represent *vague* linguistic statements whereas the input fuzzy sets reflect the uncertainty associated with the *imprecise* measurement process, and these two quantities are generally distinct. A fuzzy input distribution effectively low pass filters or averages neighbouring outputs and as the width of the input set grows (increasingly imprecise measurements), a greater emphasis is placed on neighbouring output values and the system becomes more conservative in its recommendations [19].

Defuzzification

When a fuzzy output set $\mu_B(y)$ is formed as the output of the inferencing process, it is necessary to compress this distribution to produce a single value, representing the output of the fuzzy system. This process is known as defuzzification and currently there are several implementation methods. The two most common are the mean of maximum and the centre of gravity algorithms. These can be classed as truncation and algebraic defuzzification methods, respectively, as the former bases the output estimate on only one piece of information (or at most an average of several) because the output is the value which has the largest membership in $\mu_B(y)$, whereas the latter uses the normalised weighted contribution from every point in the output distribution. The centre of gravity defuzzification algorithm tends to give a smoother output surface as there is a more gradual transition between the rules as the input is varied.

The centre of gravity defuzzification process is defined by:

$$y(\xi) = \frac{\int_Y \mu_B(y)\, y\, dy}{\int_Y \mu_B(y)\, dy} \tag{4.1.8}$$

and an important insight into how fuzzy systems process information can be made when it is assumed that the input is represented as a singleton fuzzy set, algebraic truncation operators and B-spline membership functions are used. The summation operator for representing the union means that each fuzzy set can be defuzzified separately, and using the product implication operator allows the input and output terms in the fuzzy rules to be analysed independently. Also using normalised rule confidence vectors and B-spline fuzzy membership functions together with the product intersection operator reduces the fuzzy output calculation to a simple linear combination of the multivariate fuzzy input sets, as given by:

$$y(\xi) = \sum_{i=1}^{h} \mu_{A^i}(\xi)\, \theta_i \qquad (4.1.9)$$

Therefore the shape of the fuzzy output surface is determined by the form of the fuzzy input sets. Each weight represents an estimate of the output for that particular fuzzy input set and is given by defuzzifying the information contained in the corresponding rule confidence vector. The modelling and generalisation of these fuzzy systems depend on the type of fuzzy input sets used and is completely *decoupled* from the output representation.

4.1.8 Weights and rule confidences

It is important to understand clearly the relationship between the weights used in Eq. (4.1.9) and the fuzzy rule confidences stored in the knowledge base. This then allows a comparison to be made with various neural networks and the class of neurofuzzy systems to be defined, as will be described in the following section.

Several authors have independently derived very similar results to that described above, except that often the full implications of the fuzzy representation have not been considered completely. For instance, consider when the rule confidence matrix is *binary*: i.e. for each fuzzy input set one and only one rule confidence is non-zero and that takes a value of one. This restriction means that the weights can only be one of a finite number of values (the "centres" of the fuzzy output sets). Allowing more than one rule confidence to be active, for each fuzzy input membership function, with confidences that lie in the unit interval, allows the corresponding weight to assume any value that lies the supports of the fuzzy output sets.

In addition, when the fuzzy output sets are defined by symmetrical B-spline basis functions of order k (≥ 2), the following relationship holds:

$$\theta_i = \sum_{j=1}^{q} c_{ij} y_j^c = c_i^T y^c \qquad (4.1.10)$$

where y_j^c is the centre of the jth output set, there exist q fuzzy output sets and the rule confidences are defined by:

$$c_{ij} = \mu_{B^j}(\theta_i) \qquad (4.1.11)$$

i.e. there exists an *invertible* mapping between a weight and the corresponding fuzzy rule confidence vector. Knowledge can be represented in either form and no information is lost when it is transformed. The mapping between a weight and the rule confidence vector for B-splines of order 2 is shown in Fig. 4.7 and this process can be regarded as *fuzzifying* a weight and *defuzzifying* a single rule confidence vector, and when symmetric B-splines of order k (≥ 2) are used no information is lost in this process.

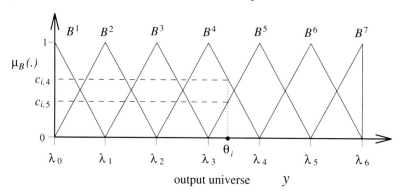

Figure 4.7: The relationship between a weight and the corresponding rule confidence vector.

4.2 Neurofuzzy Systems

As described in chapter 3, there are many different neural network architectures, but one important class of systems can be initialised and explain its actions using fuzzy rules. These networks are called *neurofuzzy* systems as they combine aspects of both neural networks (their structure and learning algorithms) and fuzzy logic (the vague, informative statements and expressions) in a single, nonlinear information processing device. Gaussian Radial Basis Function (GRBF) neural networks [159] can be classed as neurofuzzy systems, as well as the piecewise polynomial B-spline networks shown in Fig. 3.3. Another important neurofuzzy system is called the Mixture of Experts Model (MEM) where a locally "optimal" functional approximation is combined with other local mappings using a normalised weighted procedure, where the weightings are state dependent and represent the *possibility* of a local expert being correct.

4.2.1 B-spline neurofuzzy systems

When the fuzzy membership functions are defined using univariate B-splines of order k, a centre of gravity defuzzification algorithm is used, algebraic fuzzy reasoning operators are employed and the input is represented as a singleton fuzzy set, the output of the fuzzy system is given in Eq.(4.1.9). In this expression, the multivariate fuzzy sets $\mu_{A^i}(\xi) \equiv \phi_i(\xi)$ are just multivariate B-spline basis functions defined on an n-dimensional lattice (a complete rule base) and the weight vector θ is the linear set of adjustable parameters. Therefore this fuzzy system can be mapped directly onto the architecture shown in Fig. 3.3, and its modelling and generalisation abilities depend on the form and distribution of the B-spline basis functions.

This insight has many important consequences both for implementing fuzzy systems and for analysing their performance theoretically. It is an almost trivial insight to say that these neurofuzzy systems are *universal approximators*, as a fuzzy system produces piecewise polynomial mappings and by the Stone-Weierstrass theorem, the set of polynomial functions are universal approximators. It is also worth noting that increasing either the number of the B-spline fuzzy sets *or* the order of the splines, makes it possible to prove this result. This interpretation is also important because, as is described in section 4.3, the set of adjustable parameters (weights) is *linear* and there exists a lot of standard learning theory that can be applied to analyse the behaviour of these adaptive systems [19]. It also means that there is a considerable reduction in the computational cost of the system when it is implemented as described in Eq. (4.1.9) rather than performing the full fuzzification $\rightarrow$ inference $\rightarrow$ defuzzification calculation. Finally, and perhaps most importantly, this interpretation allows more advanced neurofuzzy learning algorithms to be developed which can exploit redundancy in the training data and be applied to high-dimensional modelling and control problems, and this is discussed further in section 4.3.2.

4.2.2 Gaussian radial basis function neurofuzzy systems

When the fuzzy system is implemented as described previously, but with membership functions represented using univariate Gaussian functions, the fuzzy network is structurally identical to a *normalised* GRBF algorithm, where the multivariate fuzzy input sets form a partition of unity, as described in Eq.(3.4.5). This modified GRBF network has a normalisation factor introduced by the denominator term in the original centre of gravity defuzzification algorithm which ensures that the quality of the output surface is not affected by variations in the strength of the internal representation formed by the network. It is interesting to note that the mean of maxima defuzzification process also normalises the fuzzy system's output as it is only affected by the relative height of the distribution and not the actual membership value.

Interpreting the multivariate fuzzy input membership functions as normalised Gaussian mappings has one important advantage in that it can be used to overcome the restriction that the basis functions' centres must be defined on an n-dimensional lattice. Various supervised and unsupervised learning rules can be used to select, optimise and cluster the fuzzy sets' centres, and therefore this representation is extremely useful when the data (training *and* testing) lie in a restricted part of the input space [159]. Gaussian functions are also *infinitely* differentiable and hence the fuzzy approximation and its derivatives of any order can be estimated (whether the model is trained accurately enough is another question though). Another important property is due to the local (but not strictly compact) support of the Gaussian basis functions. Normalising the output can make their supports appear global when only one basis function significantly contributes to the output, but almost compact when other basis function are defined nearly. Hence the basic form is very flexible. A truly compact support, infinitely differentiable Gaussian basis function has also been proposed by Werntges [165].

The Gaussian mapping is the only Radial Basis Function (RBF) which can be written as a product of univariate Gaussian functions, hence no other RBF network can be represented as a neurofuzzy system.

4.2.3   Mixture of experts neurofuzzy systems

MEM neurofuzzy systems are characterised by the fact that instead of associating a weight that represents an estimate of true output with each fuzzy input set, an n-dimensional *function* is stored instead [145]. These functions represent a set of local models, or *experts*, and the fuzzy input sets are used to weight the contribution made by each expert, reflecting the possibility in that model being correct. Therefore, the output of an MEM neurofuzzy system can be expressed as:

$$y(\xi) = \sum_{i=1}^{h} \mu_{A^i}(\xi)\, \theta_i(\xi) \qquad (4.2.1)$$

where $\theta_i(\xi)$ is now a function rather than a single value. Generally, each $\theta_i(\xi)$ is a linear mapping:

$$\theta_i = \theta_{i,0} + \theta_{i,1}\xi_1 + \cdots + \theta_{i,n}\xi_n \qquad (4.2.2)$$

and the fuzzy input sets can be represented using B-splines, Gaussian basis functions or as a multi-layer, sigmoidal decomposition of the input space as proposed by Jacobs and Jordan [74]. Storing linear functions rather than a single weight means it is a lot easier to approximate many smooth function despite the memory requirements being n times as great. For instance, when a GRBF neurofuzzy network tries to approximate a function which is locally linear, the width of the fuzzy sets must be extremely large for the approximation error to be small. This in turn means that the optimisation calculation for the weights is extremely ill-conditioned. However, the MEM neurofuzzy systems could easily model this type of mapping as the local expert models are linear and the fuzzy input sets form a partition of unity.

4.3   Adaptive Fuzzy Models

There are many ways of adapting the fuzzy systems which are described in this chapter, although the insights produced when algebraic fuzzy operators are coupled with centre of area defuzzification and singleton fuzzification strategies, allow a thorough theoretical framework to be derived which can predict the performance of these adaptive neurofuzzy networks. It is postulated that the best parameters to train are those for which there exists a clear and direct (transparent) relationship between them and the system's output. This not only improves the condition of the learning process, it also simplifies the validation procedures as it is easier to understand the modifications made to the adaptive neurofuzzy systems.

The three-layered architecture for the neurofuzzy systems shown in Fig. 3.3 and described by Eq.(4.1.9), illustrates how the parameters can be separated into two distinct categories, where the output is *linearly* (the weights) or *nonlinearly* (membership functions) dependent on them. Adjusting the set of linear parameters is much easier than training the nonlinear ones, and there are many learning algorithms that can be used for which convergence can be proven and rates of convergence estimated. Hence, the weights (rule confidences) are the easiest parameters to adapt, they are completely transparent as they represent estimates of the output for each fuzzy input set and so when a neurofuzzy system is iteratively trained *on-line*, the weight vector is the most convenient set of parameters to adapt. In order to adapt the *shape* of the fuzzy input sets and possibly change the structure by introducing/deleting membership functions, a complex nonlinear training procedure must be used which can only, realistically, be performed off-line. This section describes several training rules for updating the neurofuzzy systems, both on-line and off-line, and discusses their performance.

4.3.1 On-line rule adaptation

As has been discussed in chapters 2 and 3, iterative learning procedures can be used to adapt the *linear* parameter (weight) vector of a neurofuzzy system. As data are received which provide instantaneous examples of the system's desired behaviour, the weight vector can be iteratively adjusted in order to improve its performance. The Least Mean Square (LMS) and the Normalised Least Mean Square (NLMS) learning algorithms are based on performing gradient descent on an instantaneous (highly singular) Mean Square Error (MSE) performance surface. In general, the performance measure is given by:

$$J(k) = \epsilon^2(k) = (\hat{y}(k) - y(k))^2 \tag{4.3.1}$$

where $\epsilon(k)$ is the output error, and this produces learning rules of the form:

$$\text{LMS}: \quad \Delta\theta(k) = \delta\epsilon(k)\phi(k) \tag{4.3.2}$$

$$\text{NLMS}: \quad \Delta\theta(k) = \delta\epsilon(k)\frac{\phi(k)}{\|\phi(k)\|_2^2} \tag{4.3.3}$$

where $\phi(k)$ is the vector composed of the set of outputs of the fuzzy input sets i.e. $\phi_i(k) = \mu_{A^i}(\xi(k))$, δ is the learning rate and $\Delta\theta(k) = \theta(k+1) - \theta(k)$ is the change in the weight vector at time k.

These training procedures update the weight vector parallel to the unit transformed input vector $\phi(k)/\|\phi(k)\|^2$ for a step length of:

$$\text{LMS}: \quad \delta\epsilon(k)\,\|\phi(k)\|^2 \qquad\qquad (4.3.4)$$

$$\text{NLMS}: \quad \delta\frac{\epsilon(k)}{\|\phi(k)\|^2} \qquad\qquad (4.3.5)$$

Therefore both learning algorithm update their weight vector along the same search direction, it is only the distance travelled along it that differs. They base the calculation of both the search direction and the step length on just one training datum. However, it is not possible to estimate which part of the output error is due to an error in the weight vector and that due to measurement or modelling error. Similarly, successive search directions are highly correlated when two inputs are close, relative to the input lattice. So, despite the obvious simplicity of this training rule, its performance can be degraded when there exists significant modelling and measurement error and successive training samples are similar.

Such objections can be partially overcome by using a set of higher order learning algorithms that assess the current system's performance using more than one training datum. This information can then be used to introduce some orthogonality into the search directions and to partially filter out any measurement or modelling error. There are a wealth of optimisation techniques available for solving such a problem when the desired mapping is *static*, however, the main reason for using on-line learning is to cope with time-varying plants. Also these standard algorithms are generally too expensive to be implemented in real-time. Therefore, it is worthwhile considering how a set of computationally inexpensive, linear optimisation algorithms can be derived for modelling time-varying mappings [19, 113].

Central to the problem of deriving higher-order learning algorithms is the ability to construct a representative *store* of training data. The store is a fixed size, and should be composed of training data which are both recent (in order to model time-varying plants) and sufficiently exciting (partially uncorrelated to increase the rate of parameter convergence). There are a variety of ways for updating but one of the simplest is to match the current training datum with all those in the store and to discard the oldest, similar one, where the time and spatial matches are weighted concepts. Given this extra information, many learning rules can be developed that generalise the classical LMS-type algorithms. The one which has been found to work best in practice sets the search direction to be parallel to the transformed input vector in the store which currently has a maximum error residual and selects the step length such that it minimises the current MSE over all the data. This orthogonalises the search direction and dampens the steps taken along it [3, 19].

Adapting the rule confidence

This section has investigated how the weights in a neurofuzzy system may be trained efficiently, neglecting the "fuzzy" rule confidence representation. This can be justified because there exists a *learning equivalence*[4] between the weights and rule confidences, as shown in section 4.1.8. Therefore a neurofuzzy system can be implemented and trained in weight space, then the learnt function can be explained using fuzzy rules, due to the invertible mapping between weights and rule confidences. However, it is possible to train the rule confidences directly, and this section discusses how this is achieved and whether it is desirable.

The output of a neurofuzzy system is *linearly* dependent on the rule confidence matrix as:

$$y(\xi) = \sum_{i=1}^{h} \mu_{A^i}(\xi) \left(\sum_{j=1}^{q} c_{ij} \, y_j^c \right) \tag{4.3.6}$$

which can be rewritten as:

$$y(\xi) = \phi^T(\xi) \, C \, y^c \tag{4.3.7}$$

where C is the rule confidence matrix of size $(h \times q)$. Both the fuzzy input sets and the output set centres are static, hence the network's output is a linear function of the rule confidences in the space defined by:

$$A \otimes y^c \tag{4.3.8}$$

where A is a matrix whose k^{th} row is defined by $\phi^T(\xi(k))$. This space is *inherently* singular as the output set centres do not depend on the network's input. Therefore there are an infinite number of optimal rule confidence matrices which produce the optimal system output. Even when the rule confidences are restricted to lie in the unit interval and each rule confidence vector is normalised (sums to unity), the optimisation problem is still not uniquely posed. Therefore, gradient descent procedures converge to an answer but it does not necessarily have a sensible linguistic interpretation.

In place of the desired output in the LMS and NLMS learning rules it is possible to use a *desired rule confidence vector*, which is obtained by evaluating the output sets' memberships for the desired output in a manner analogous to that described in section 4.1.8. A rule confidence vector error can then be formed by subtracting the current (weighted average) estimate from the desired one and using this quantity in a fuzzy LMS or NLMS rule confidence updating algorithm:

$$\Delta c_i(k) = \delta \left(\hat{c}(k) - \sum_j \mu_{A^j}(\xi(k)) \, c_j(k) \right) \mu_{A^i}(\xi(k)) \tag{4.3.9}$$

where $\hat{c}(k)$ is the desired rule confidence vector which can be formed from the membership of the desired output in the fuzzy output sets. An adaptive neurofuzzy system which

---

[4]The input/output behaviour of the two adaptive systems is the same, it is only their internal representations that differ.

is trained using this learning rule is learning equivalent to an analogous weight-based network whose weight vector is adapted using the LMS algorithm.

To summarise, equivalent fuzzy rule confidence training algorithms can be derived for neurofuzzy systems, but their implementation cost is greater and they are less transparent due to the inherent singularity of the optimisation problem.

4.3.2   Off-line network structuring

Determining the *structure* of a neurofuzzy system is a complex nonlinear, iterative optimisation problem that must find out which inputs are important, whether a single network can be replaced by two or more simpler ones and what is the internal representation (number and shape of the fuzzy sets) formed by each (sub)network. This is an extremely important task in any modelling procedure because when the network's structure is inappropriate it will either be overparameterised and start to model noise or else it will not be flexible enough to store the required information. In addition, forming an appropriate internal representation endows the composite system with the following desirable attributes:

- **improved** generalisation abilities;

- **reduced** amount of required training data;

- **better** network conditioning; and

- **simplified** network structure leading to a more transparent rule base.

This section describes how this problem may be addressed by exploiting additive and local redundancy, as described in section 3.6.

The problems conventional neurofuzzy networks have in approximating functions defined on medium to high-dimensional input spaces may not be immediately apparent from the preceeding description, but it is a serious problem when complex real-world applications are developed. For instance, consider the resources required by a neurofuzzy system in modelling a 5-dimensional mapping with 7 univariate fuzzy sets defined on each (input and output) axis: $7^6 \approx 120,000$ rule confidences or $7^5 \approx 17,000$ weights. In either case, at least $20,000$ well distributed training data are required for the system to generalise correctly across the whole input space and in many practical situations, it is simply not possible to collect even a fraction of this information. This problem is commonly known as the *curse of dimensionality* (see section 3.6).

Global redundancy

One of the simplest methods for alleviating the curse of dimensionality which occurs in conventional neurofuzzy systems is to try and exploit any global redundancy which occurs in the desired function. The simplest ways of achieving this is only to include those inputs which contribute a significant amount of information in the model. Reducing the size of the input space by ignoring redundant inputs can simplify the neurofuzzy system's structure by orders of magnitude. Another technique for simplifying the system's structure is to exploit any *additive* redundancy [14, 19, 78]. In the simplest case, the network could be composed of a linear combination of nonlinear univariate models (ANOVA) of the form:

$$y(\xi) = \sum_{i=1}^{n} s_i(\xi_i) \qquad (4.3.10)$$

where $s_i(.)$ is a univariate, nonlinear neurofuzzy subnetwork. The overall network is extremely simple and its computational requirements increase *linearly* in n, but it can only model an additive interaction between the different input variables i.e. no cross product polynomial terms of the form $\xi_1 \xi_2$. A more general model is of the form:

$$y(\xi) = \sum_{i=1}^{r} s_i(\xi_i) \qquad (4.3.11)$$

where r is the number of (possibly multivariate) neurofuzzy subnetworks whose input vectors ξ_i lie in a smaller dimensional subspace. This approach tries to approximate the underlying desired function using a linear combination of smaller dimensional neurofuzzy subnetworks, as shown in section 3.6.

Additive dependencies can be identified in an iterative fashion by having a very simple model to begin with (possibly empty) and a set of univariate networks in a store. Each univariate network in the store can be considered as possible univariate subnetwork in the neurofuzzy network, and the one which reduces the current output error the most (weighted according to its complexity) is included in the neurofuzzy network. Note that these networks can be trained using standard optimisation techniques such as conjugate gradient [138]. Because the additive redundancy is being explicitly modelled, the condition of the optimisation problem is improved. This procedure iteratively constructs neurofuzzy networks that are formed from a linear combination of univariate nonlinear subnetworks. Higher dimensional subnetworks can also be found by allowing a tensor product operation between a subnetwork currently in the model with another univariate network from the store or else one contained in the current model. This increases the dimension of the original subnetwork by one and means that the new subnetwork can store exactly the information contained in the previous two, as well as being able to model new cross product terms. In practice, at any time it is possible either to include a new univariate network or else form a tensor product operation, and the one which is the most *statistically significant* (i.e. reduces an error based complexity function the most) is included in the current model.

Local partitioning

In developing an automatic network construction algorithm it is also necessary to devise a procedure that can optimise the number and position of the fuzzy membership functions. The previously described inductive learning procedure determines which input variables are significant in accurately predicting the plant's performance but it is also generally necessary to refine (increase, modify and/or simplify) a subnetwork's structure. However, the development of a well-defined iterative refinement procedure depends on the representation used by each of the subnetworks. The simplest assumption to make is that the multivariate fuzzy sets are defined on a lattice (taking the tensor product of each of the univariate fuzzy subnetworks). Under this assumption, each of the univariate fuzzy subnetworks can be refined individually and a new multivariate subnetwork formed from their tensor product. When B-spline fuzzy membership functions are used, a new knot can be inserted in the knot vector (halfway between two previous values for instance) and this introduces a new basis function and locally modifies k of the previous membership functions. Figure 4.8 shows such a case, where a new B-spline fuzzy membership function (knot) is inserted in a univariate subnetwork. This process locally modifies the existing fuzzy membership functions and adds an extra degree of freedom (another piecewise linear segment) in the model. Inserting a single knot in the univariate basis functions is equiva-

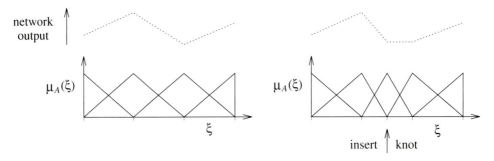

Figure 4.8: Inserting a new B-spline fuzzy membership function (knot).

lent to adding a new line on a 2-dimensional lattice or an $(n-1)$-dimensional hyperplane on an n-dimensional input space. By restricting the new knot values to occur halfway between neighbouring values and evaluating every possible position, it is possible to determine which refinement produces the best fit to the unmodelled data, that subnetwork can be modified and the whole network retrained.

Sometimes however, a subnetwork can become too flexible as a one step ahead inductive learning procedure is not always optimal and mistakes can be made. An input may be incorrectly included in the model or because the knot insertion procedure only chooses its candidate values from the set of interval midpoints, an overly complex subnetwork may be formed. Therefore, network pruning is an essential tool in limiting its complexity. Knots (fuzzy membership functions) can be deleted and subnetworks split up into smaller dimensional ones, and if either of these procedures a model which fits the data as well as the more complex one, the overall neurofuzzy network should be simplified [146].

Finally, it is worthwhile considering the limitations of the above scheme. Most neuro-fuzzy systems have a set of basis functions that are defined on a lattice. Even though the above procedure can partially overcome this limitation by modelling any global additive redundancy explicitly, each subnetwork is still defined on a smaller dimensional lattice. More complex input space partitioning techniques can be used to partially overcome this restriction, although most of these which can be interpreted as fuzzy rules still rely on splitting the input space parallel to an axis, as is shown in Fig. 4.9. Even by allowing for

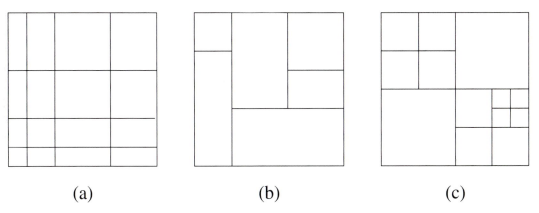

(a)　　　　　　　　　　(b)　　　　　　　　　　(c)

Figure 4.9: A lattice (a), tree (b) and a hierarchical (b) partitioning of the input space.

these partitioning strategies which can exploit local redundancy, it may be necessary to rotate and decorrelate inputs prior to presenting them to each subnetwork. With these possible restrictions in mind, the basic network structuring algorithm can produce good parsimonious medium scale models for otherwise high-dimensional fuzzy systems.

4.4　Adaptive Fuzzy Control Systems

This chapter has so far described the basic theory behind implementing a fuzzy algorithm within a fuzzy system, and discussed the implications of using different membership functions, operators etc. However, these learning fuzzy systems must be used as *part* of an overall design scheme when they are used for adaptive control. This section reviews several different direct and indirect adaptive fuzzy controllers, thus providing alternative approaches from those described in [65]. To begin with, the basic problem of plant modelling is considered.

4.4.1   Adaptive fuzzy modelling

It is relatively simple to use a fuzzy system as an adaptive plant model, because the desired training signal is directly available (the measured plant's output) and all that needs to be determined is the structure of the fuzzy system. The learning process is then a linear optimisation procedure and convergence to a global minimum can be shown to occur when the data satisfies boundedness constraints. For the simplest structure shown in Fig. 4.10a, the input to the fuzzy plant model is formed from time delayed measurements

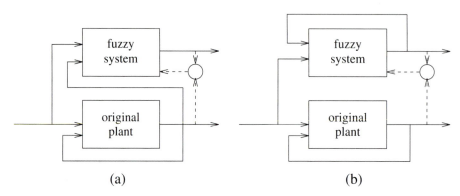

(a)   (b)

Figure 4.10: Two basic fuzzy plant models. The training signal path is shown as a dashed line.

of the control input and the plant's output. In this case the training data is assumed to be generated by a process described by:

$$y(k) = f(\xi(k)) \tag{4.4.1}$$

where $f(.)$ is the unknown nonlinear mapping that characterises the unknown plant and $\xi(k) = (y(k-1), y(k-2), \ldots, y(k-n_y), u(k-1), \ldots, u(k-n_u))$ is the plant's information vector. A fuzzy model can then be formed as:

$$\hat{y}(k) = \hat{f}(\xi(k)) \tag{4.4.2}$$

where $\hat{f}(.)$ represents the fuzzy system. The input vector is composed of the plant's output values and control inputs. Given that the measurements are correct, the orders n_y and n_u have been correctly estimated and that the control input is persistently exciting, the fuzzy system will satisfactorily approximate $f(.)$. Even when these conditions are not satisfied, the fuzzy system can be shown to track the plant's output i.e.,

$$\lim_{k \to \infty} \epsilon(k) = 0 \tag{4.4.3}$$

although the network will be unable to generalise appropriately.

When the measured output is corrupted by noise, this modelling procedure can be biased, as the inputs are corrupted and it cannot be determined whether an output error is due to parameter mismatch or because the input is incorrect. To overcome this difficulty, the output of the model can be used as its own input instead of the measured value (assuming that the fuzzy system's output is sufficiently close to the true value) producing a fuzzy model of the form:

$$\widehat{y}(k) = \widehat{f}(\xi(k)) \tag{4.4.4}$$

where $\xi(k) = (\widehat{y}(k-1), \widehat{y}(k-2), \ldots, \widehat{y}(k-n_y), u(k-1), \ldots, u(k-n_u))$ is the fuzzy system's information vector, as shown in Fig. 4.10b. The learning rules for training these *recurrent* fuzzy models are nonlinear, as the current tracking error may be influenced by parameter errors from several time steps in the past, rather than the most recent one.

These fuzzy systems can be used to model an expert operator who knows how to control a process that is difficult to model (e.g. cement kilns), or to directly construct an inverse model of the process or, more directly, to predict the states of the unknown process which is subsequently used in a predictive control algorithm or an indirect controller design. The success of all of these applications depends on the structure of the fuzzy rule bases (estimating the quantities n_y and n_u and the form of the nonlinear mapping) and the algorithms described in section 4.3.2 can be used for this purpose.

4.4.2 Indirect adaptive fuzzy control

Adaptive control algorithms can be broadly separated into direct and indirect schemes [12]. In the latter approach, an explicit model of the controller is never formed and the adaptive fuzzy model is used to construct a control signal *indirectly*. There has been a large number of indirect adaptive fuzzy control systems developed within the last 5 years (previously every adaptive fuzzy system was direct) and this section briefly describes some of these approaches, attempting to emphasise their common features.

Affine systems

A substantial amount of work has been aimed at developing control algorithms for a class of systems called *affine*, which are linear in the control signal and can therefore be expressed as:

$$y(k) = f(\xi(k)) + g(\xi(k))u(k-1) \tag{4.4.5}$$

where $f(.)$ and $g(.)$ are nonlinear functions and $\xi(k) = (y(k-1)\ldots, y(k-n_y), u(k-2), \ldots, u(k-n_u))$ This *additive* and *multiplicative* decomposition of the original plant into two submodels where one is multiplied by the current control signal can easily be represented in a fuzzy system which has two or more subnetworks where one depends on $u(k-1)$ that is modelled using two B-splines of order 2 (i.e. linear on a compact domain). Therefore the previously described network structuring algorithm can be used to determine exactly which inputs are important and how they should be represented, and the fuzzy plant model is given by:

$$\hat{y}(k) = \hat{f}(\xi(k)) + \hat{g}(\xi(k))u(k-1) \tag{4.4.6}$$

Given that certain conditions are satisfied by f and g, such as smoothness and g being bounded away from zero and knowing that the fuzzy subnetworks can approximate f and g sufficiently closely, it is possible to design a controller based on the *certainty equivalence* principles. Therefore, let the desired output at time k be denoted by $y_m(k)$, then the control signal is calculated from:

$$u(k-1) = \frac{y_m(k) - \hat{f}(\xi(k)) + d^T \varepsilon(k)}{\hat{g}(\xi(k))} \tag{4.4.7}$$

where d is a vector such that the roots of the polynomial $q^n + d_1 q^{n-1} + \cdots + d_n$ all lie in the unit circle and $\varepsilon(k) = (\epsilon(k-1), \ldots, \epsilon(k-n))$ is a vector of delayed output errors. It is possible to design simple Lyapunov updating rules for the fuzzy subnetworks $\hat{f}$ and $\hat{g}$ such that system stability can be proven [159]. Strictly speaking, this also requires the existence of a stabilising supervisory controller which activates whenever the inputs lie outside the domain of the fuzzy system, similar to the system described in chapter 5.

One of the most appealing features of affine systems is that their structures can be determined using input-output linearisation techniques, and this gives an upper bound on the size of the input space.

Fuzzy model inversion

Instead of assuming an affine structure for the plant model, it may be possible to reduce this to monoticity in the control input and to invert the fuzzy plant model either directly by inverting the fuzzy rule base or indirectly by a numerical search procedure or by fitting a known function to the fuzzy model which can then be easily inverted. In all of these algorithms, a fuzzy model is used to map the current plant's state and control inputs to the next state. Given that a reference model can supply a desired behaviour, the rule base must be inverted to find $u(k-1)$ given that $y(k) = y_m(k)$ is known. The first technique, which inverts the rule base directly by exploiting the facts that the input and output variables are both represented using fuzzy sets and the implication operator is invertible, may seem attractive because the controller is formed from an explicit set of (inverted) fuzzy rules. However, this approach has poor generalisation properties as the (inverted) rule confidence vectors are no longer normalised (sum to unity) and the control system is unable to extrapolate information properly at the input domain's boundary.

Inverting the fuzzy rule base indirectly using a numerical search procedure is probably the most accurate method, although it can be time consuming and there may exist discontinuities in the mapping (due to uninitialised fuzzy rule confidences), hence search routines which assume certain smoothness properties cannot be used. One technique which has proven to be effective is to fit a (globally) linear function to the mapping $u(k-1) \to y(k)$, in a mean squared error sense, and then invert this linear relationship [114]. However, this implicitly assumes that there exists an affine structure in the plant and if this is true, the techniques mentioned in the previous section could be more appropriate.

Other approaches

Many other techniques can use a fuzzy model as part of their control design procedure, such as d-step-ahead predictive control algorithms which use optimisation algorithms to search for a sequence of control signals that minimise a pre-defined performance function. This procedure does not explicitly use the linguistic properties of a fuzzy system, but the localised learning coupled with an efficient implementation could be sufficient to justify its use. Similarly, a technique is described in chapter 5 which uses the fuzzy systems to model the (nonlinear) gains of a linear system and then a controller design is performed using the locally linear model. Here the fuzzy systems are used because they are linear in their adjustable parameters and the optimisation procedure is well-conditioned. It may be possible to initialise both of these systems using expert, fuzzy knowledge about the process, and although this is a useful feature, it is the network's learning abilities which make it appropriate for these tasks.

4.4.3 Direct adaptive fuzzy control

A direct adaptive fuzzy controller builds a fuzzy controller directly. The fuzzy controller maps current state and desired state information to a recommended control signal, and one of the simplest (and most popular) measures the error (current deviation from a set point) and the error rate, and based on this input signal the fuzzy system outputs either the control signal or a recommended change in the control, as illustrated in Fig. 4.11. The direct fuzzy controller implements a *nonlinear* control algorithm that can be initialised and validated in a natural manner using vague, linguistic rules. Indeed, the majority of the fuzzy control systems that have been developed in the Far East are *static*, and possibly the only reason why a fuzzy system is preferable is that common sense, fuzzy rules can be implemented and iteratively modified in a natural manner. Direct adaptive fuzzy controllers attempt to acquire these rules by observing how a plant reacts to various control actions, and update their strategy accordingly. Therefore, direct (and indirect) adaptive fuzzy controllers are simply nonlinear learning systems and they should be subject to the same rigorous convergence, stability and robustness analysis as conventional techniques.

Most direct adaptive controllers are error based; each input signal is based on the error between the current plant's output and a demanded set point (such as its derivative and/or integral); thus mimicking a standard nonlinear Proportional + Integral + Derivative (PID) controller. However, one of the main reasons for using direct adaptive fuzzy systems is to control *nonlinear* plants and unless special precautions are taken, such as altering the

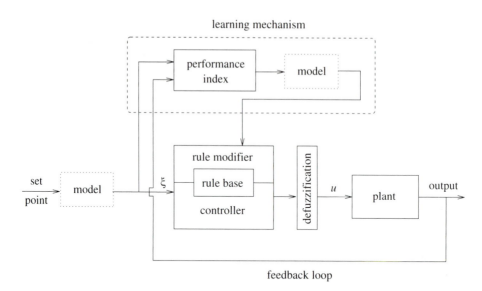

Figure 4.11: A direct fuzzy control system.

definition of the fuzzy sets according to the operating point, the plant's local dynamical behaviour and hence the required error based controller mapping will depend on the operating point. This procedure is equivalent to gain scheduling for linear controllers as altering the definition of the fuzzy sets effectively alters the overall gain of the system.

Learning mechanisms

In building direct adaptive control systems, the fundamental question that must be answered is: which part of the output error is due to an incorrect control signal? For instance, using raw step response data to adapt the controller is infeasible because that would be equivalent to demanding that the plant should achieve instantaneous transitions in the system's state. Therefore this signal must be filtered (in some sense) using a coarse model of the process. This may be placed before the controller in order to make the demanded plant output signal realisable, or else this type of knowledge may be implicitly encoded in the learning mechanism. It is not true to say that these direct adaptive fuzzy controllers do not need a model of the process, rather some kind of prior knowledge about its dynamical behaviour (either quantitative or qualitative) must be available, and the performance of the overall control system *will* depend on the quality of this specification.

The original self-organising fuzzy controllers used a performance index, together with an estimate of the plant's inverse Jacobian [130]. The performance index is used to relate the plant's output measurements (such as deviation from a set point and the current rate of change of error) to a signal which characterises the system's performance. A model of the plant's inverse Jacobian is then used to map these performance measures to the required changes in the control signals that are necessary to improve the system's performance. This is shown in Fig. 4.11 except that the first plant model, which transforms the demanded step response, is not used. This procedure has been shown to work well, and some of the limitations (and possible solutions) of this basic approach are discussed in [178].

Using a performance index and an (inverse) model of the process is certainly not the only approach for training direct adaptive fuzzy controllers, although in much of the literature on this subject there has been a desire to keep the training mechanisms fuzzy. In contrast, we believe that a lot of progress can be made by examining more conventional learning algorithms for which convergence, stability and robustness results exist, and sometimes they are easier to extend to the multivariate case. For instance, if the plant's output depends linearly on the control signal, but nonlinearly on other state and time-delayed control inputs, direct learning algorithms can be developed [159] that are similar to those traditionally used in the adaptive control literature [12]. Fuzzy controllers are simply nonlinear, numeric mappings that can be initialised and validated using linguistic fuzzy rules. The linguistic representation is only important as an interface between the controller's knowledge base and the designer, whereas the performance (rate of convergence) of the adaptive fuzzy controller depends on the structure of the fuzzy system and the learning mechanism; both of which should be chosen to make the overall adaptation process as simple as possible. Whether or not this uses traditional "fuzzy" concepts (min/max operators, performance indices) is immaterial.

Chapter 5

Neural Network and Fuzzy Logic Based Adaptive Control

5.1 Introduction

In the design and implementation of nonlinear control systems, *linearisation* is one of the most widely used techniques. Linearisation around a fixed operating point produces a locally linear model for which a controller can be synthetised by conventional linear controller design methods, [51]. However, any change in the system operation condition will lead to changes in the parameters of the linearised model, causing a deterioration of the performance of the closed loop system. To avoid this, the parameters of the controller need to be modified in response to variations in the plant characteristics and in operating point. The performance of the closed loop system will depend upon these parameter changes and satisfactory performance will depend upon accurate representation of these parameter variation. If the relationships between the parameters of linear model and operating points are known exactly and the operating points are measurable then the parameter changes are completely specified. Often, these relationships are complicated, unknown, nonlinear functions but at least for some systems the operating points are measurable or may be inferred. Examples of such dynamic systems are hydraulic rigs, AC motors, electrical power plant control systems, [172], and hydraulic turbine systems [150]. For instance, in power plant control the measurable operating points are voltage, power and speed at the terminal of the generators, [172]. Similarly, for aero gas turbines, system performance (thrust) is typically a function of attitude and mach number which are measurable operating points. For these kind of nonlinear system, if the linear model at fixed operating points is called a local system representation, then global system representation can be realized via a set of parameterised linear models corresponding to the different operating points. The total number of linear model may be finite or infinite. That is, one can globally express certain nonlinear systems as a set of linear systems whose parameters are unknown nonlinear functions of its measurable operating points. If the general form of single-input and single-output (SISO) nonlinear systems is given by

$$y(k) = \quad f(y(k-1), y(k-2), \cdots, y(k-n), u(k-d),$$
$$u(k-d-1), \cdots, u(k-d-m)) \tag{5.1.1}$$

where $f(.)$ is an unknown nonlinear input/output function, $y(k)$ is the measured output of the system, $u(k)$ is the measured input to the system, the integers n and m are known a priori or assumed and represent the orders of the system and d, is the known time delay of the system, then the system discussed in the previous paragraph can be expressed as

$$y(k) = \quad a_1(O_k)y(k-1) + \cdots + a_n(O_k)y(k-n) +$$
$$b_0(O_k)u(k-d) + b_1(O_k)u(k-d-1) + \cdots +$$
$$b_m(O_k)u(k-d-m) \tag{5.1.2}$$

where $a_i(O_k)$ and $b_j(O_k)$ $(i = 1, 2, ..., n; \; j = 0, 1, ..., m)$ are the unknown parameters which are functions of the measured operating points O_k, which are p dimensional vectors. It can be seen that system (5.1.2) is a special case of system (5.1.1) when the measurable operation points O_k depend upon the past values of the system input and output. However, if we assume that the nonlinear function $f(.)$ is first order differentiable with respect to its arguments $\{y(k-1), y(k-2), ..., y(k-n), u(k-d), ..., u(k-d-m)\}$, then local linearisation can always be performed which will produce a system of the form (5.1.2), where the input and output are actually of incremental value type and the measurable operating points $O_k = \{y(k-1), y(k-2), ..., y(k-n), u(k-d), u(k-d-1)..., u(k-d-m)\}$. Also, if the nonlinear function $f(.)$ in system (5.1.1) can be expressed as $f(.) = \sum_i f_i(y(k-i)) + \sum_j f_{n+j+1}(u(k-d-j))$ and all $f_i(.), f_j(.)$ are first order differentiable, then it can be expressed by system (5.1.2) as well. Moreover, it can be seen that the systems described by Eq. (5.1.2) are more general than those expressed by Eq. (2.5.13)-(2.5.16).

   For example, in a turbofan engine, the spool speed/fuel flow response in a full flight envelop operation is characterized by a nonlinear relation of the form

$$N = f(WF, t_N, K_N, P, T) \tag{5.1.3}$$

where N is the spool speed, WF is engine fuel flow, t_N is the spool time constant, K_N is spool steady-state gain, P stands for pressure and T represents temperature. About an operating point, the response can be closely approximated by a first order dynamic model

$$\Delta N = \frac{K_N}{1 + t_N s} \Delta WF \tag{5.1.4}$$

where the parameters t_N and K_N are nonlinear functions of altitude A and mach number M, i.e.,

$$t_N = f_1(A, M) \tag{5.1.5}$$
$$K_N = f_2(A, M) \tag{5.1.6}$$

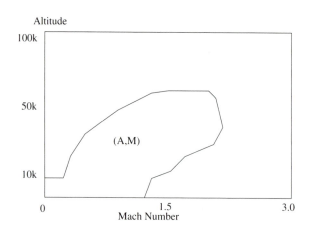

Figure 5.1: A full flight envelop.

where the range of (A, M) for a full flight envelop is shown in Fig. 5.1.

The discrete version of (5.1.4) becomes

$$\Delta N(k) = a_1(A, M)\Delta N(k - 1) + b_1(A, M)\Delta WF(k - 1) \tag{5.1.7}$$

Let $y(k) = \Delta N(k)$, $u(k) = \Delta WF(k)$ and $O_k = (A, M)$, the system of the form (5.1.2) is, again, obtained.

Research into the identification and adaptive control of the system (5.1.2) has been carried out since 1980. Early studies, [62, 112, 173], simply replaced the measurable operating points O_k by time k and the identification of the time-varying parameters was mainly performed via the modified recursive least square (RLS) algorithms and the variance matrix was modified when changes in the system parameters were detected thorough residual signals, [62]. Later, under the assumption of smoothness about the unknown nonlinear functions $a_i(O_k)$ and $b_j(O_k)$, Taylor series were used to expand these nonlinear functions in terms of either the time k, [173], or the measurable operating points P_k, [150]. The identification of the unknown functions is then performed by identifying the constants of the Taylor series. Using the estimated models, adaptive controllers can be designed based upon the *certainty equivalence* principle, such as d-step-ahead adaptive control and poles and zeros assignment adaptive control. The stability of closed loop system can be established by using the *technical lemma* discussed in chapter 2. In these early approaches, [62, 112], the opportunity to speed up the convergence of estimation by making use of measurable information was neglected due to simple design scheme adopted. As for the Taylor series expansion based approaches, [150] and [173], the high order smoothness conditions for the unknown nonlinear functions are too restrictive for practical systems and only low order Taylor series expansions can be used, [150]; resulting in low accuracy models.

It is well known that neural networks can approximate any continuous nonlinear function defined on a compact set to an arbitrary accuracy, therefore, if neural networks are used to approximate the unknown nonlinear functions in the system (5.1.2) and online training is employed to adjust the weights of the neural networks, estimates for the unknown parameters $a_i(O_k)$ and $b_j(O_k)$ can be obtained. The controller can then be designed using the standard procedures employed in linear adaptive control. Compared with the most widely used nonlinear ARMA model (5.1.1), system (5.1.2) can be modelled using less complicated neural networks, since the number of input variables to each neural network approximating $a_i(O_k)$ or $b_i(O_k)$ is usually less than $n+m+1$. If the dimension of O_k is much less than $n + m + 1$, it is expected that the total number of the weights used by the neural networks modelling system (5.1.2) will be smaller than that used for modelling the system (5.1.1), since for some neural networks implementations, such as Radial Basis Functions (RBF), [55] and [24], the total number of the weights increases exponentially with the number of the inputs to the neural network. This fact will lead to the simplified design for system (5.1.2) and will, of course, reduce the computational burden for the training procedure during which a large number of the weights are updated.

In this chapter, a recently developed ANN based modelling and control for the system (5.1.2) will be described. The modelling and control of two different types of systems were considered by Wang, Harris and Brown, [154]. The first system assumed that the nonlinearities $a_i(O_k)$ and $b_j(O_k)$ can be exactly modelled by the neural networks whereas the second system only required that these nonlinearities can be approximated by the neural networks to a pre-specified accuracy. The first group of systems was referred to as *matching systems* and the second group of systems was defined as *mismatching systems*. For mismatching systems, model uncertainty could exist and the modified recursive least square algorithm was used to ensure both weight convergence and closed loop system stability.

Problems may also arise when neural networks are used to estimate the parameters of system (5.1.2) because these networks are defined on a bounded input subspace. This means that measurable operating points should lie in a uniformly bounded set. However, this is only true for some systems and in general the following three situations may occur:

- Measurable operating point is independent of the system input and output;

- Measurable operating point is a subset of the system input and output;

- Measurable operating point is composed of both the independent variables and a subset of the system input and output.

For the first case, one can assume the uniform boundness of measurable operating points and a compact set can then be chosen on which the basis functions are defined [16]. This is the main problem that this chapter will address, and it will be seen later that both convergent training and the stable control of the closed loop system is achieved. The second case is also discussed, where a sliding control strategy, [134], is proposed to ensure that the input and output signals are bounded, which means that they will lie inside a compact set. The third case is then only a combination of the first two cases.

5.2   Matching Systems

5.2.1   Identification for unknown matching systems

As has been defined in section 5.1, the nonlinearities in a matching system can be exactly modelled by neural networks. Let

$$N_{ai}^*(O_k) = \sum_{q=1}^{l} w_{iq}^a \phi_q(O_k) \qquad (5.2.1)$$

$$N_{bj}^*(O_k) = \sum_{q=1}^{l} w_{jq}^b \phi_q(O_k) \qquad (5.2.2)$$

be the neural networks which reproduce the nonlinearities $a_i(O_k)$ and $b_j(O_k)$, where w_{ik}^a and w_{jk}^b are the weights of the neural network and $\phi_k, (k = 1, 2, \cdots, l)$ are the known basis functions for either B-splines (see chapter 3) or CMAC neural networks or even radial basis function (RBF) with pre-specified centers and l is a known integer. The architecture of these networks, which has been given in Fig. 3.3, is modified to the one shown in Fig. 5.2.

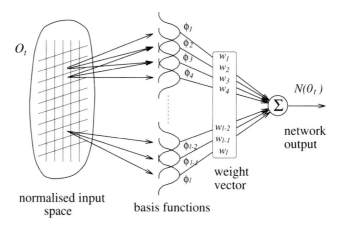

Figure 5.2: Neural network approximation of $a_i(O_k)$ and $b_j(O_k)$.

Using N_{ai}^* and N_{bj}^*, the original system (5.1.2) can be expressed by

$$y(k) \;=\; N_{a1}^*(O_k)y(k-1) + N_{a2}^*(O_k)y(k-2) + \cdots$$

$$N_{an}^*(O_k)y(k-n) + N_{b0}^*(O_k)u(k-d) + \cdots$$
$$N_{bm}^*(O_k)u(k-d-m) \tag{5.2.3}$$

which can then be transferred into the following form

$$y(k) = \theta^T \Phi(k-1) \tag{5.2.4}$$

$$\theta = [w_{11}^a, w_{12}^a, \cdots, w_{1l}^a, w_{21}^a, w_{22}^a, \cdots w_{n1}^a, w_{n2}^a, \cdots, w_{nl}^a$$
$$w_{01}^b, w_{02}^b, \cdots, w_{0l}^b, \cdots, w_{m1}^b, \cdots, w_{ml}^b]^T \in R^{l(n+m+1)} \tag{5.2.5}$$

$$\Phi(k-1) = [\phi_1 y(k-1), \phi_2 y(k-1), \cdots \phi_l y(k-1), \phi_1 y(k-2),$$
$$\phi_2 y(k-2), \cdots \phi_l y(k-2), \cdots \phi_1 u(k-d), \phi_2 u(k-d), \cdots,$$
$$\phi_l u(k-d) \cdots \phi_1 u(k-d-m), \phi_2 u(k-d-m), \cdots,$$
$$\phi_l u(k-d-m)]^T \in R^{l(n+m+1)} \tag{5.2.6}$$

It can be seen that Eq. (5.2.4) has exactly the same form as the widely used linear regression equation in system identification, where the system output $y(k)$ is linear in terms of the unknown parameter vector θ. Therefore, the same recursive algorithms can be used to estimate the unknown parameter vector θ. Let $\hat{\theta}(k)$ be the estimate of θ at time t, then the normalized back-propagation algorithm studied in [128] takes the following form

$$\Delta\hat{\theta}(k) = \hat{\theta}(k) - \hat{\theta}(k-1) = \eta \frac{\Phi(k-1)\varepsilon(k)}{c + \Phi^T(k-1)\Phi(k-1)} \tag{5.2.7}$$

$$\varepsilon(k) = y(k) - \hat{y}(k-1) \tag{5.2.8}$$

$$\hat{y}(k-1) = \hat{\theta}^T(k-1)\Phi(k-1) \tag{5.2.9}$$

where $\eta \in (0, 2)$, c is positive constant which can be arbitrarily small and $\hat{\theta}(0)$ is an initial value. The theorem below summarizes the properties of this algorithm.

Theorem 5.1 *When the normalized back-propagation algorithm (5.2.7) - (5.2.9) is applied to the data set $\{y(k), u(k)\}$ generated by original matching systems, the equalities:*

$$\lim_{k\to\infty} \frac{\varepsilon^2(k)}{c + \Phi^T(k-1)\Phi(k-1)} = 0 \tag{5.2.10}$$

$$\lim_{k\to\infty} \| \hat{\theta}(k) - \hat{\theta}(k-k_0) \| = 0 \tag{5.2.11}$$

hold, where k_0 is a positive integer. Furthermore, if $\| \Phi(k-1) \| < \infty$, then $\lim_{k\to\infty}(y(k) - \hat{y}(k-1)) = 0$

The proof of this theorem can be found in [128] and is omitted here. If we further denote

$$\hat{\theta}_{ai}(k) = [\hat{w}_{i1}^a, \hat{w}_{i2}^a, \cdots \hat{w}_{il}^a]^T \in R^l \tag{5.2.12}$$

$$\hat{\theta}_{bj}(k) = [\hat{w}^b_{j1}, \hat{w}^b_{j2}, \cdots, \hat{w}^b_{jl}]^T \in R^l \tag{5.2.13}$$

$$\phi(k-1) = [\phi_1(O_k), \phi_2(O_k), \cdots, \phi_l(O_k)]^T \in R^l \tag{5.2.14}$$

$$\xi(k-1) = [y(k-1), y(k-2), \cdots, y(k-n), u(k-d), \cdots,$$
$$u(k-d-m)]^T \in R^{n+m+1} \tag{5.2.15}$$

$$i = 1, 2, ..., n; \; j = 0, 1, 2, ..., m \tag{5.2.16}$$

then the algorithm (5.2.7)-(5.2.9) can be realized in parallel as follows

$$\hat{\theta}_{ai}(k) = \hat{\theta}_{ai}(k-1) + \eta \frac{\phi(k-1)y(k-i)\varepsilon(k)}{c + \phi^T(k-1)\phi(k-1)\xi^T(k-1)\xi(k-1)} \tag{5.2.17}$$

$$\hat{\theta}_{bj}(k) = \hat{\theta}_{bj}(k-1) + \eta \frac{\phi(k-1)u(k-d-j)\varepsilon(k)}{c + \phi^T(k-1)\phi(k-1)\xi^T(k-1)\xi(k-1)} \tag{5.2.18}$$

$$i = 1, 2, ..., n; \; j = 0, 1, 2, ..., m \tag{5.2.19}$$

where the only coupled part of the algorithm is the modelling error $\varepsilon(k)$ (sometimes called predictive error in adaptive control) calculated from Eqs. (5.2.8)-(5.2.9). That is, at each stage, all the previous subvectors $\hat{\theta}_{ai}(k-1)$ and $\hat{\theta}_{bi}(k-1)$ are needed to form $\hat{\theta}(k-1)$ which is used together with $\phi(k-1)$ and $\xi(k-1)$ to produce $\hat{y}(k-1)$ and $\varepsilon(k)$. In Fig. 5.3, the structure of the networks representing the estimated system is shown.

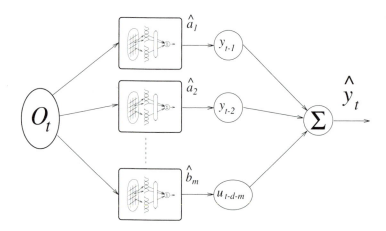

Figure 5.3: Neural network model for system (5.1.2).

It can be seen from this figure that the network consists of two hidden layers. The weights in the first layer (next to the input layer) are those updated by the above algorithm and the weights in the second layer simply take the time-delayed values of the input and output $\{y(k)\}$ and $\{u(k)\}$. With $\hat{\theta}(k)$, the parameter estimates for $a_i(O_k)$ and $b_j(O_k)$

become

$$\hat{a}_i(O_k) = \sum_{k=1}^{l} \hat{w}_{ik}^a(k-1)\phi_k(O_k) \tag{5.2.20}$$

$$\hat{b}_j(O_k) = \sum_{k=1}^{l} \hat{w}_{jk}^b(k-1)\phi_k(O_k) \tag{5.2.21}$$

$$i = 1, 2, ..., n; \qquad j = 0, 1, ..., m$$

and the estimated model for system (5.1.2) can be formed by

$$
\begin{aligned}
y(k) = & \ \hat{a}_1(O_k)y(k-1) + \hat{a}_2(O_k)y(k-2) + \cdots + \hat{a}_n(O_k)y(k-n) \\
& \ \hat{b}_0(O_k)u(k-d) + \hat{b}_1(O_k)u(k-d-1) + \cdots + \hat{b}_m(O_k)u(k-d-m)
\end{aligned}
\tag{5.2.22}
$$

5.2.2 d-Step-ahead predictive control

Using the estimates of the weights and parameters, on-line control strategies can be readily formed which realize various requirements on the closed loop system. In this section, a d-step-ahead predictive control strategy for system (5.1.2) will be described. For this purpose, let us define

$$\hat{A}(k, q^{-1}) = 1 + \sum_{i=1}^{n} \hat{a}_i(O_k)q^{-i} \tag{5.2.23}$$

$$\hat{B}(k, q^{-1}) = \sum_{j=0}^{m} \hat{b}_j(O_k)q^{-j} \tag{5.2.24}$$

where q^{-1} is a unit backward shift operator introduced in chapter 2, then the estimated model for system (5.1.2) can be expressed by

$$\hat{A}(k, q^{-1})y(k) = \hat{B}(k, q^{-1})q^{-d}u(k) \tag{5.2.25}$$

By solving the Diophantine equation

$$1 = \hat{F}(k, q^{-1})\hat{A}(k, q^{-1}) + \hat{G}(k, q^{-1})q^{-d} \tag{5.2.26}$$

the two polynomials $\hat{F}(k, q^{-1})$ and $\hat{G}(k, q^{-1})$, of orders $d-1$ and $n-1$ respectively, can be obtained. Left multiplying both sides of Eq. (5.2.25) by $\hat{F}(k, q^{-1})$ and using the equality (5.2.26), the *estimated model* for system (5.1.2) can be rewritten as

$$y(k) = \hat{G}(k, q^{-1})y(k-d) + \hat{E}(k, q^{-1})u(k-d) \tag{5.2.27}$$

$$\hat{E}(k, q^{-1}) = \hat{F}(k, q^{-1})\hat{B}(k, q^{-1}) \tag{5.2.28}$$

Let $\hat{y}(k)$ be the predict value of the system output $y(k)$, the d-step-ahead predictor

$$\hat{y}(k) = \hat{G}(k, q^{-1})y(k - d) + \hat{E}(k, q^{-1})u(k - d) \tag{5.2.29}$$

can be directly obtained. As a result, the d-step-ahead predictive control signal $u(k)$ which makes $y_r(k) = \hat{y}(k)$ satisfies

$$y_r(k + d) = \hat{G}(k, q^{-1})y(k) + \hat{E}(k, q^{-1})u(k) \tag{5.2.30}$$

where $y_r(k)$ is the desired value of the output of system (5.1.2). The stability of the closed loop system is established using the following theorem.

Theorem 5.2 *Assume that system (5.1.2) in invertibly stable (i.e. the bounded desired output $y_r(k)$ can be realized by the bounded control signal) and that the basis functions $\phi_k(O_k), (k = 1, 2, ..., l)$ are uniformly bounded, then if the control signal $u(k)$ of Eq. (5.2.30) is applied to system (5.1.2), the resulting closed loop system is stable, i.e.,*

$$|y(k)| < +\infty; \quad |u(k)| < +\infty. \tag{5.2.31}$$

Proof: Define

$$A(k, q^{-1}) \quad = \quad 1 + \sum_{i=1}^{n} a_i(O_k)q^{-i} \tag{5.2.32}$$

$$B(k, q^{-1}) \quad = \quad \sum_{j=0}^{m} b_j(O_k)q^{j} \tag{5.2.33}$$

as the polynomial operators for unknown system (5.1.2), then following the same procedure as in Eqs. (5.2.26)-(5.2.28), it can be seen that system (5.1.2) can also be expressed in its d-step-ahead predictive form

$$y(k) = G(k, q^{-1})y(k - d) + E(k, q^{-1})u(k - d) \tag{5.2.34}$$

and again, the orders of $G(k, q^{-1})$ and $E(k, q^{-1})$ are the same as those of $\hat{G}(k, q^{-1})$ and $\hat{E}(k, q^{-1})$. Comparing this expression with Eq. (5.3.29), it can be shown that

$$y(k) - \hat{y}(k - 1) = \varepsilon(k) \tag{5.2.35}$$

Since the d-step-ahead predictive control law $u(k)$ defined in Eq. (5.2.30) makes $y_r(k) = \hat{y}(k)$, it follows that

$$y(k) - y_r(k - 1) = \varepsilon(k) \tag{5.2.36}$$

holds for the closed loop system. This is an important result which will be used to prove the stability of the closed loop system. Since the basis functions $\phi_k(.), k = 1, 2, ..., l$ are uniformly bounded, there exist a positive constant c_0 such that

$$|\phi_k(O_k)| \le c_0; \quad (k = 1, 2, ..., l) \qquad \forall O_k \tag{5.2.37}$$

Due to the fact that

$$\Phi^T(k-1)\Phi(k-1) = \phi^T(k-1)\phi(k-1)\xi^T(k-1)\xi(k-1) \tag{5.2.38}$$

there exists a positive constant c_1 such that

$$\| \Phi(k-1) \| \le c_1 \| \xi(k-1) \| \tag{5.2.39}$$

Using Eq. (5.2.36) and the fact that system (5.1.2) is invertibly stable, it can be shown that

$$\| \xi(k-1) \| \le c_2 + c_3 \max_k | \varepsilon(k) | \tag{5.2.40}$$

and therefore

$$\| \Phi(k-1) \| \le c_4 + c_5 \max_k | \varepsilon(k) | \tag{5.2.41}$$

where c_2, c_3, c_4 and c_5 are positive constants. By making use of the first result in theorem 5.1 and the technical lemma in chapter 2, it can be concluded that

$$\lim_{k\to\infty} \varepsilon(k) = 0 \tag{5.2.42}$$

Thus from inequality (5.2.40), it can be seen that

$$\| \xi(k-1) \| < \infty \tag{5.2.43}$$

which means (see Eq. (5.2.15)) that both $u(k)$ and $y(k)$ are uniformly bounded. □□

5.3 Mismatching Systems

In systems where there exists model mismatch, it is assumed that there is a known positive number δ such that the inequalities

$$|N_{ai}^*(O_k) - a_i(O_k)| \le \delta \tag{5.3.1}$$
$$|N_{bj}^*(O_k) - b_j(O_k)| \le \delta \tag{5.3.2}$$

will hold uniformly for $i = 1, 2, ..., n$; $j = 0, 1, ..., m$ and for all O_k. In this case, equation (5.2.3) can be modified to the following form

$$\begin{aligned}
y(k) = {}& N_{a1}^*(O_k)y(k-1) + N_{a2}^*(O_k)y(k-2) + \cdots + \\
& N_{an}^*(O_k)y(k-n) + N_{b0}^*(O_k)u(k-d) + \cdots + \\
& N_{bm}^*(O_k)u(k-d-m) + \Delta f(\xi(k-1))
\end{aligned} \tag{5.3.3}$$

where $\xi(k-1)$ is defined the same as that in section 5.2 and

$$\begin{aligned}
\Delta f(\xi(k-1)) = {}& (N_{a1}^* - a_1)y(k-1) + (N_{a2}^* - a_2)y(k-2) + \cdots \\
& (N_{an}^* - a_n)y(k-n) + (N_{b0}^* - b_0)u(k-d) + \cdots \\
& + (N_{bm}^* - b_m)u(k-d-m)
\end{aligned} \tag{5.3.4}$$

Note here that N_{ai}^*, N_{bj}^*, a_i and b_j are used to stand for $N_{ai}^*(O_k)$, $N_{bj}^*(O_k)$, $a_i(O_k)$ and $b_j(O_k)$ for simplicity.

Using Eqs. (5.3.1)-(5.3.2) it can be shown that

$$|\Delta f(\xi(k-1))| \leq (n+m+1)\delta \, \| \, \xi(k-1) \, \| \tag{5.3.5}$$

Therefore, using neural networks, the original unknown system (5.1.2) can be expressed as

$$y(k) = \theta^T \Phi(k-1) + \Delta f(\xi(k-1)) \tag{5.3.6}$$

where θ and $\Phi(k-1)$ are defined in Eqs. (5.2.5)-(5.2.6). Due to the existence of the nonlinear uncertainty $\Delta f(\xi(k-1))$, the identification and control algorithm developed in previous section has to be modified to ensure that the closed loop system is stable. This is achieved by introducing both a normalized signal and a dead-zone to the algorithm which allows the weight convergence and the stability of the closed loop system to be studied. Like the assumption made in section 5.2, the operating points O_k are still independent of the system input and output and lie in a compact set, on which the basis function $\phi_k(k = 1, 2, ..., l)$ are defined.

5.3.1   Identification

Define the normalized signal $\rho(k)$ as

$$\rho(k) = \alpha\rho(k-1) + (1-\alpha)\max(1, \|\xi(k-1)\|) \tag{5.3.7}$$

where $\alpha \in (0, 1)$ and $\rho(0) \neq 0$, denote

$$y_\rho(k) = y(k)/\rho(k) \tag{5.3.8}$$
$$\Phi_\rho(k-1) = \Phi(k-1)/\rho(k) \tag{5.3.9}$$
$$\Delta_\rho f(\xi(k-1)) = \Delta f(\xi(k-1))/\rho(k) \tag{5.3.10}$$

then the normalized equation

$$y_\rho(k) = \theta^T \Phi_\rho(k-1) + \Delta_\rho f(\xi(k-1)) \tag{5.3.11}$$

can be obtained with

$$\|\Delta_\rho f(\xi(k-1))\| \leq (n+m+1)\delta = \delta_1 \tag{5.3.12}$$

Based on these two equations, the following modified recursive least square algorithm can be obtained.

$$\Delta\hat{\theta}(k) = \hat{\theta}(k) - \hat{\theta}(k-1) = \frac{\eta(k)P(k-2)\Phi_\rho(k-1)\varepsilon_\rho(k)}{1 + \Phi_\rho^T(k-1)P(k-2)\Phi_\rho(k-1)} \tag{5.3.13}$$
$$\varepsilon_\rho(k) = y_\rho(k) - \hat{y}_\rho(k-1) \tag{5.3.14}$$
$$\hat{y}_\rho(k-1) = \hat{\theta}^T(k-1)\Phi_\rho(k-1) \tag{5.3.15}$$
$$P^{-1}(k-1) = P^{-1}(k-2) + \eta(k)\Phi_\rho(k-1)\Phi_\rho^T(k-1) \tag{5.3.16}$$

where

$$\eta(k) = \begin{cases} 1 & |\varepsilon_\rho(k)| \geq \nu(k) \\ 0 & |\varepsilon_\rho(k)| < \nu(k) \end{cases}$$

(5.3.17)

and

$$\nu(k) = \frac{\delta_1}{(1 + \Phi_\rho^T(k-1)P(k-2)\Phi_\rho(k-1))^{1/2}}$$

(5.3.18)

and $\hat{\theta}(0)$ and $P(0)$ are initial values. Again, this identification algorithm can be performed in parallel and in this case the variance matrix $P(k-2)$ is a block diagonal matrix. The convergence of the above algorithm is proved by following theorem.

Theorem 5.3 *When the algorithm (5.3.13)-(5.3.18) is applied to the input and output data $\{u(k), y(k)\}$ of system (5.1.2), then*

$$\lim_{k \to \infty} \frac{\varepsilon_\rho^2(k)}{1 + \Phi_\rho^T(k-1)P(k-2)\Phi_\rho(k-1)} = \delta_1^2$$

(5.3.19)

$$\lim_{k \to \infty} \left\| \hat{\theta}(k) - \hat{\theta}(k - k_0) \right\| = 0$$

(5.3.20)

where k_0 is also a positive integer.

Proof: Following the same line as that in [57], it can be shown that with the Lyapunov function

$$V(k) = \left(\hat{\theta}^T(k) - \theta^T \right) P^{-1}(k-1) \left(\hat{\theta}(k) - \theta \right)$$

(5.3.21)

the equality

$$V(k) - V(k-1) = \begin{cases} -\frac{\varepsilon_\rho^2(k)}{1 + \Phi_\rho^T(k-1)P(k-2)\Phi_\rho(k-1)} + \delta_1^2 & |\varepsilon_\rho(k)| \geq \nu(k) \\ 0 & |\varepsilon_\rho(k)| < \nu(k) \end{cases}$$

(5.3.22)

holds. Therefore $\Delta V(k) = V(k) - V(k-1)$ is non-positive which means that $V(k)$ is a lower bounded and non-increasing series, hence

$$\lim_{k \to \infty} \Delta V(k) = 0$$

(5.3.23)

exists. It can then be concluded that both results in the theorem hold. □□

Using $\hat{\theta}(k)$, the d-step-predictive model similar to that in section 5.2 becomes

$$\hat{y}(k) = \hat{G}(k, q^{-1})y(k-d) + \hat{E}(k, q^{-1})u(k-d)$$

(5.3.24)

where polynomial operators $\hat{G}(k, q^{-1})$ and $\hat{E}(k, q^{-1})$ are the same as in Eqs. (5.2.27)-(5.2.28) and the only difference here is that the parameter estimate $\hat{\theta}(k)$ is calculated by the algorithm (5.3.13)-(5.3.18) rather than by Eqs. (5.2.7)-(5.2.9).

5.3.2 Controller design

Due to the inaccuracy of the neural network approximations to the actual nonlinear parameters for mismatching systems, d-step-ahead predictive control strategy discussed before no longer provide a stable closed loop system. Therefore, in this subsection, a pole-zero assignment algorithm for mismatching system will be described. It can be seen later that such a control scheme will garantee the closed loop stability when δ is sufficiently small.

The controller in this case is constructed as follows ([86])

$$y_r(k+d) = \hat{y}(k+d) + Q(k, q^{-1})v(k) \tag{5.3.25}$$
$$u(k) = L(k, q^{-1})v(k) \tag{5.3.26}$$

where

$$Q(k, q^{-1}) = \sum_{j=0}^{n_Q} Q_j q^{-j} \tag{5.3.27}$$

$$L(k, q^{-1}) = \sum_{j=0}^{n_L} L_j q^{-j} \tag{5.3.28}$$

are the two polynomial operators determined on-line by solving

$$(1 - \hat{G}(k, q^{-1})q^{-d})Q(k, q^{-1}) + \hat{E}(k, q^{-1})L(k, q^{-1}) = T(q^{-1}) \tag{5.3.29}$$

and $T(q^{-1}) = \sum_{j=0}^{n_T} T_j q^{-j}$ is a pre-chosen stable polynomial. With this control algorithm, the following theorem ([39]) can be obtained.

Theorem 5.4 *If δ_1 is sufficiently small and all the basis functions are uniformly bounded, then all the variables in the closed loop system are uniformly bounded when the control algorithm (5.3.13)-(5.3.18), (5.3.25)-(5.3.26)) and (5.3.29) is applied to system (5.1.2).*

Since $\delta_1 = (n + m + 1)\delta$, this theorem states that the closed loop system will stable if the approximation accuracy of the neural networks to the nonlinear functions, $a_i(O_k)$ and $b_j(O_k)$, is high.

Proof: Since all the basis functions $\phi_k(O_k)$ are uniformly bounded, from Eq. (5.2.39), there is a positive constant c_1 such that

$$\| \Phi(k-1) \| \le c_1 \| \xi(k-1) \| \tag{5.3.30}$$

Therefore, the normalized signal vector $\Phi_\rho(k)$ satisfies

$$\|\Phi_\rho(k)\| \le c_1 \qquad \forall t \ge 0, \tag{5.3.31}$$

From theorem 5.3, it can be seen that there is a $t_0(\delta_1)$ such that when $t \geq t_0$, the following inequality

$$\frac{|\varepsilon_\rho(k)|}{(1 + \Phi_\rho^T(k-1)P(k-2)\Phi_\rho^T(k-1))^{\frac{1}{2}}} \leq \delta_1 + \delta_1 = 2\delta_1 \tag{5.3.32}$$

holds. As a result

$$|\varepsilon_\rho(k)| \leq 2\delta_1(1 + c_1^2 \|P(0)\|)^{\frac{1}{2}} \tag{5.3.33}$$

If we still denote

$$\varepsilon(k) = y(k) - \hat{y}(k-1) \tag{5.3.34}$$

then it can be shown from Eq. (5.3.33) that

$$|\varepsilon(k)| \leq 2\delta_1(1 + c_1^2 \|P(0)\|)^{\frac{1}{2}} \|\xi(k-1)\| \tag{5.3.35}$$

Let $\hat{G}_0(k, q^{-1}) = (1 - \hat{G}(k, q^{-1})q^{-d})$, then it can be seen that

$$[\hat{G}_0(k, q^{-1})Q(k, q^{-1}) + \hat{E}(k, q^{-1})L(k, q^{-1})]v(k) = \\ \hat{G}_0(k, q^{-1})y_r(k+d) - \hat{G}(k, q^{-1})\varepsilon(k) \tag{5.3.36}$$

Since Eq. (5.2.29) holds and $T(q^{-1})$ is stable, there exist a positive number δ_2 such that

$$|v(k)| \leq 2\delta_1\delta_2(1 + c_1^2 \|P(0)\|)^{\frac{1}{2}} \|\xi(k-1)\| + \delta_3 \tag{5.3.37}$$

where $0 < \delta_3 < \infty$. Using Eq. (5.3.26), it can be shown that there are positive constants δ_4, δ_5 and δ_6 such that

$$|u(k-d+1)| \leq 2\delta_1\delta_4\delta_2(1 + c_1^2 \|P(0)\|)^{\frac{1}{2}} \|\xi(k-d)\| + \delta_3\delta_4 \tag{5.3.38}$$

$$|y(k)| \leq |\varepsilon(k)| + |\hat{y}(k-1)| \leq 2\delta_1(1 + c_1^2 \|P(0)\|)^{\frac{1}{2}} \|\xi(k-1)\| + \\ 2\delta_1\delta_5(1 + c_1^2 \|P(0)\|)^{\frac{1}{2}} \|\xi(k-d-2)\| + \delta_6 \tag{5.3.39}$$

and in general

$$\|\xi(k)\| \leq 2\delta_1 c_6 \sum_{i=0}^{n-1} \|\xi(k-d-2-i)\| + 2\delta_1 c_6 \sum_{j=0}^{m-1} \|\xi(k-d-j)\| + \\ 2\delta_1 c_6 \sum_{h=0}^{n-1} \|\xi(k-1-h)\| + c_7 \tag{5.3.40}$$

is satisfied, where

$$c_6 = (1 + c_1^2 \|P(0)\|)^{\frac{1}{2}} \max(\delta_4\delta_2, \delta_5, 1) \tag{5.3.41}$$
$$c_7 = n\delta_6 + (m+1)\delta_3\delta_4 \tag{5.3.42}$$

are bounded positive numbers which depend on the parameter estimates and the choice of polynomial operator $T(q^{-1})$. Therefore, from Eq. (5.3.40), $\|\xi(k)\|$ will be uniformly bounded if

$$c_6\delta_1 < \frac{1}{6} \qquad (5.3.43)$$

This is equivalent to the statement that all the variables of the closed loop system will be uniformly bounded if δ_1 is sufficiently small.   □□

To summarize, the following algorithm can be used to obtain stable closed loop control of nonlinear mismatching systems:

i) Set the initial values of the estimates $\hat{\theta}(0)$ and $P(0)$;

ii) Use the algorithm (5.3.13)-(5.3.18) to calculate the parameter estimates $\hat{\theta}(k)$;

iii) Form polynomial operators $\hat{G}(k, q^{-1})$, $\hat{E}(k, q^{-1})$ and solve Eq. (5.3.29) for $Q(k, q^{-1})$ and $L(k, q^{-1})$;

iv) Calculate the control signal $u(k)$ by using Eqs. (5.3.25)-(5.3.26).

The structure of the closed loop system is shown in Fig. 5.4.

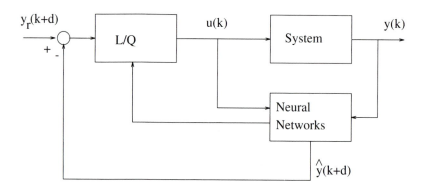

Figure 5.4: The neural network adaptive closed loop control system.

5.3.3 Examples and discussions

To investigate the applicability of the proposed modelling method in section 5.2 and this section, the following system is considered

$$y(k) = 2O_1 e^{-10(O_1^2 + O_2^2)} y(k-1) + 0.5sin(O_1^2 + O_2^2)y(k-2) + 0.6u(k) \qquad (5.3.44)$$

where $O_k = (O_1, O_2)^T$ is the measurable operating point, $y(k)$ is the measured output sequence and $u(k)$ is the input sequence. During the simulation, O_1, and O_2 are uniformly distributed random sequences in the interval $[-0.5, 0.5] \times [-0.5, 0.5]$, and $u(k)$ obeys a normal distribution $N(0, 1)$. In practice the operating points will vary in a structured and progressive manner, frequently locally or piecewise linear, through demands made by the plant operator. Two, 2-dimensional, second order B-spline ANNs are used for modelling and the corresponding knot vectors are given by

$$\Lambda = \{-0.5, -0.4, -0.3, -0.2, -0.1, 0.0, 0.1, 0.2, 0.3, 0.4, 0.5\} \tag{5.3.45}$$

for both O_1 and O_2. The partition of the input domain is shown in Fig.5.5.

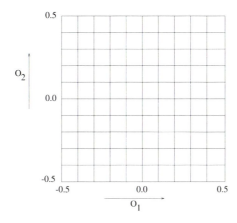

Figure 5.5: The partitioning of the input space.

The mismatching system is considered with the accuracy bound given by $\delta = 0.0005$. Therefore δ_1 in Eq. (5.3.12) becomes 0.015. The parameter α in Eq. (5.3.7) is set as 0.43 and $\rho(0) = 0.35$. The initial variance matrix for the learning algorithm (5.3.13)-(5.3.18) is given by $P(0) = 10^5 I_{243}$ and the initial values of the weights are zero.

With 3000 training data points and 10 iterations, both the true nonlinear surfaces and the estimated surfaces are given in Fig. 5.6. There exists 121 adjustable weights in each neural network, and it can be clearly seen from these figures that good approximations are obtained.

Based on Eqs. (5.3.25)-(5.3.26), a pole-assignment controller for the nonlinear process (5.3.44) is obtained by setting the desired polynomial as

$$T(q^{-1}) = 1 - 1.5q^{-1} + 0.56q^{-2} \tag{5.3.46}$$

and the desired output $y_r(k)$ in this case is a square wave with a magnitude of 1.0 and a period of 200 sample times. Moreover, in order to track the desired output, a factor of

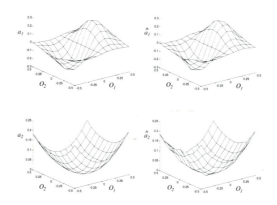

Figure 5.6: Estimation results.

$1-q^{-1}$ is included in the polynomial $Q(k, q^{-1})$. The parameters in $L(k, q^{-1})$ and $Q(k, q^{-1})$ are updated by solving Eq. (5.3.29) and the controller's output is then generated via Eqs. (5.3.25) and (5.3.26). The closed loop response following the above modelling procedure after 2,600 training/sample periods is shown in Fig 5.7 where the remaining 400 training data points are plotted. It can be readily seen that both desirable learning and good transient response is achieved despite continual random variations in the operating points. Interperiod fluctuations in the controlled response are caused by the random variations in the operating points, whereas in practice this would be a smooth transitional process. Nevertheless the controlled response is sufficient for all but the most demanding processes.

5.4   General Systems

In the above sections, it is assumed that the measurable operating points are independent of the system input and output variables and that a compact support for the neural networks exist. It is also based on these assumptions that convergent parameter estimations and stable controllers are obtained. However, as has been discussed in section 5.1, some practical systems exist in which the measurable operating points are dependent on the system input and output variables. We call this kind of system the general system and one example of which is the generator control of power system, where the varying parameters are the unknown nonlinear functions of the frequency and voltage at the terminal of the generators. Therefore, the algorithms developed in sections 5.2-5.3 need to be modified in order to obtained stable control.

Without loss of generality, it is assumed throughout this section that all the unknown nonlinear functions $a_i(O_k)$ and $b_j(O_k)$ are bounded and that the operating point vectors

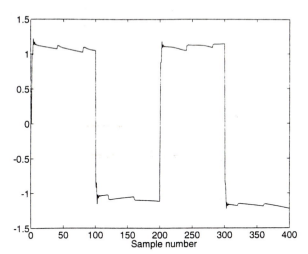

Figure 5.7: The closed loop response.

O_k are formed from a subset of the past values of the system output $y(k)$. That is,

$$O_k = [y(k-1), y(k-2), \cdots, y(k-p)] \in R^p \qquad (5.4.1)$$

where $p \leq n$ is a positive integer. Let S be a compact set in R^p such that the neural network N_{ai} and N_{bj} satisfy

$$|N_{ai}(O_k) - a_i(O_k)| \leq \varepsilon_f \qquad (5.4.2)$$
$$|N_{bj}(O_k) - b_j(O_k)| \leq \varepsilon_f \qquad (5.4.3)$$

on the set S. The control strategy in this section consists two parts which are defined by

$$u(k) = \begin{cases} u_1(k) & O_k \in S \\ u_2(k) & O_k \in R^p - S \end{cases} \qquad (5.4.4)$$

where $u_1(k)$ is the control signal generated by the control algorithms in either section 5.2 or section 5.3 and $u_2(k)$ is sliding control which is used to bring the operating point O_k from $R^p - S$ into S.

Since system (5.1.2) can be expressed in its state-space form as

$$x(k+1) = A(O_k)x(k) + B(O_k)u(k) \qquad (5.4.5)$$
$$y(k) = C(O_k)x(k) \qquad (5.4.6)$$

where the matrices $A(O_k)$, $B(O_k)$ and $C(O_k)$ are also uniformly bounded, a sliding control $u_2(k)$ [143, 137] can be shown to exist which is able to bring the state to the sliding surface

$$s(k) = 0 \qquad (5.4.7)$$

in finite time. On this sliding surface it can be shown that the system output $y(k)$ is bounded and, therefore, with proper choice of the gains in the control $u_2(k)$, the compac set S can be reached in a finite time as well. Once this happens, the control signal $u(k$ is switched to $u_1(k)$ which, as has been shown in sections 5.2-5.3, will stabilize closed loop system. Detailed formulation of the sliding control $u_2(k)$ is not discussed here, sinc it is well covered in literature, [143, 137]. For example, one can adopt the procedure o Sarpturk in 1987 to design $u_2(k)$ by using the known lower and upper bounds for the coefficient matrices $A(O_k)$, $B(O_k)$ and $C(O_k)$.

Alternatively, one can use a one-to-one mapping to transfer the non-compact definitio domain of functions $a_i(O_k)$ and $b_j(O_k)$ into an open set which is covered by a compac set ([152]). In this way, no switching control can be obtained.

5.5 Fuzzy Logic Based Control

5.5.1 Fuzzy logic representations

Similar to B-spline neural networks, fuzzy logic systems described in chapter 4 can also be used to approximate the nonlinear coefficients in the operating point dependent system of this chapter. The structure of such a representation is given in Fig. 4.1, where the fuzzy logic performs the $O_k \in \bar{O} \rightarrow a_i \in R$ or $O_k \rightarrow b_j \in R$ mapping using the following rule set

$$\text{If } (O^1 \text{ is } A_1^v) \text{ and } \cdots \text{ and } (O^p \text{ is } A_p^v), \text{ then } a_i \text{ or } b_i \text{ is } G_i^v$$

where O^j is the jth component of O_k, A_j^v and G_i^v are labels of fuzzy sets, $v = 1, 2, \cdots, l$ and l is the total number of rules. Each IF-THEN rule defines a fuzzy implication in th product space $\bar{O} \times R$. From chapter 4, it can be seen that, when center-average defuzzifie (4.1.8), algebraic sum/product inferences (4.1.5) and singleton fuzzifier (4.1.7) are used the nonlinear coefficients a_i and b_j can be expressed as

$$a_i = \frac{\sum_{v=1}^{l} \bar{a}_i^v (\prod_{q=1}^{p} \mu_{A_q^v}(O^q))}{\sum_{v=1}^{l} (\prod_{q=1}^{p} \mu_{A_q^v}(O^q))} \tag{5.5.1}$$

$$b_j = \frac{\sum_{v=1}^{l} \bar{b}_j^v (\prod_{q=1}^{p} \mu_{A_q^v}(O^q))}{\sum_{v=1}^{l} (\prod_{q=1}^{p} \mu_{A_q^v}(O^q))} \tag{5.5.2}$$

where $\bar{a}_i^v$ and $\bar{b}_j^v$ are the points at which $\mu_{G_i^v} = 1$ and $\mu_{G_j^v} = 1$. By fixing $\mu_{A_q^v}(O^q)$ in th system, it can be further obtained that

$$a_i = \sum_{v=1}^{l} \bar{a}_i^v \phi_v(O_k) \tag{5.5.3}$$

$$a_i = \sum_{v=1}^{l} \bar{a}_i^v \phi_v(O_k) \tag{5.5.4}$$

where

$$\phi_v(O_k) = \frac{\prod_{q=1}^{p} \mu_{A_q^v}(O^q)}{\sum_{v=1}^{l}(\prod_{j=1}^{p} \mu_{A_q^v}(O^q))} \tag{5.5.5}$$

Since Eqs. (5.5.3)-(5.5.4) are in the same form as Eqs. (5.2.1)-(5.2.2), the adaptive updating laws used for the tuning of the weights in section 5.2 can be directly applied to learn $\bar{a}_i^v$ and $\bar{b}_j^v$. As a result, d-step-ahead predictive control strategy of Eqs. (5.2.29)-(5.2.30) can be readily utilised. Compared with neural network based approaches in section 5.2-5.3, the advantage of using fuzzy logic is that the human knowledge on the operating points, which is normally summarised from the experience of human operators about the process, can be directly incorperated in the construction phase of fuzzy membership functions.

Using the above procedure, the estimated model for system (5.1.2) can be expressed in a form shown in Fig. 5.8, where $\hat{a}_i$ and $\hat{b}_j$ are the estimates of the unknown parameters, $y(k)$ is the output of the estimated model.

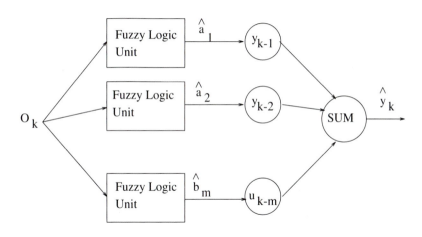

Figure 5.8: Fuzzy logic model for the system.

5.5.2 Application to MD grammage control

Figure 5.9 shows a typical papermaking system, where grammage is one of the most important quantities representing the property of the paper. Low variation of grammage will lead to uniform distribution of fibres and thus increase paper quality. The control of grammage is divided into machine direction (MD) and cross direction (CD). MD grammage of paper is mainly controlled by either increasing or decreasing the input of thick stock. However, the grammage of final products is also effected by paper moisture content, which is controlled by the steam pressure applied to the dry section of paper machines. As a result, the relationships between grammage, moisture content and thick stock flow are complex, and in most cases, they are unknown and nonlinear. Since moisture conten

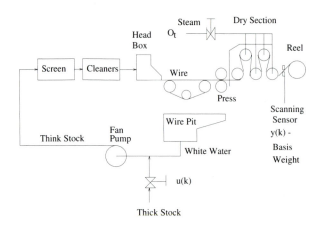

Figure 5.9: A papermaking process.

is measurable, it can be utilized as a measurable operating point, by which the system
can be expressed by the following first order equation

$$y(k) = a(O_k)y(k-1) + b(O_k)u(k-d) \qquad (5.5.6$$

where $u(k)$ is the valve opening which controls the flow of thick stock, $y(k)$ is the MI
grammage measured via scan sensors, O_k is moisture content of the paper; $a(O_k)$ and
$b(O_k)$ are nonlinear unknown coefficients of the system, $d = 5$ is the known MD time
delay which corresponds to the sampling period of 10 seconds. The linguistic variable O_k
ranging from 0.04 to 0.8, consists of ten fuzzy sets ($l = 10$) and the nominal setpoint of the
grammage is $100g/m^2$. It is assumed that $a \in R_a = [0.93, 0.98]$ and $b \in R_b = [0.01, 0.02]$
and that the range of the parameters are, again, be divided by ten fuzzy sets. Gaussian
membership function of the type in Eq. (4.1.1) are used with $\sigma_i = 0.025$ and equally
spaced centres.

To simulate process, it is assumed that the variation of moisture content is given by

$$O_k = 0.06 + 0.02sin(0.14k) + w_k \qquad (5.5.7$$

where w_k is white noise with zero mean and 0.01 variance, and that the parameters of the
process have following nonlinear nature

$$\begin{aligned}
a(O_k) &= 0.93 + 0.025cos(O_k^{2/3}) & (5.5.8 \\
b(O_k) &= 0.015 & (5.5.9
\end{aligned}$$

The d-step-ahead predictive control algorithm in section 5.2 is used and the simulation results of the closed loop step response and parameter estimation are shown in Fig. 5.10, where the setpoint change between $100g/m^2$ and $110g/m^2$ is applied. The parameter estimation results are shown in Fig. 5.11. It can be seen from these figures that the closed loop system is stable, the parameter estimation error is very small and the dynamic performance is desirable.

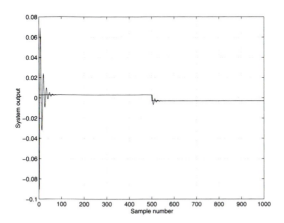

Figure 5.10: The closed loop response.

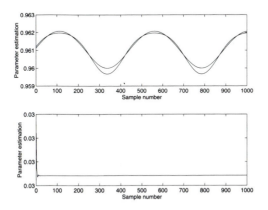

Figure 5.11: The estimation results.

5.6   Conclusions

In this chapter we have discussed the use of associative memory neural networks and
fuzzy logic of chapters 3-4 for both the modelling and the control of a class of nonlinear
systems, the parameters of which are unknown nonlinear functions of the measurable
operating points. The systems in this case can be regarded as locally linear and globally
nonlinear. Since each AMN is used to model the individual parameters, the original
system can be expressed by a two-layer neural network. The weights in the first layer are
tained either by a normalised backpropagation algorithm or by the modified recursive least
squares algorithm, whilst the weights in the output layer are simply the past values of the
system input/output. Both d-step-ahead predictive and pole-zero assignment algorithms
are discussed which have been shown to be able to stabilize the closed loop systems.
Fuzzy logic based approach is also discussed for estimating the nonlinear parameters of
the systems and its potential application to the MD grammage control of a paper machine
is illustrated through computer simulation. Desired simulation results for both AMN and
fuzzy logic based approaches have been obtained.

Many practical systems can be expressed in the form described in this chapter. Typical
examples of such systems have already been briefly outlined in section 5.1. This means
that the design procedures discussed in this chapter are very useful for the construction
of practical control systems, and we therefore end this chapter by summarising these
algorithms in the following steps:

- identify the system input, output and the measurable operating points;

- express the system in the form of (5.1.2);

- choose a suitable modelling procedure (e.g., to decide whether to use AMN or fuzzy
  logic to represent the nonlinear parameters);

- select a proper training algorithm and the structure of the controllers.

The control strategies discussed in this chapter are only used to illustrate the design
procedure. It can be seen that any adaptive control scheme can be co-operated with the
modelling part to form the controller suitable for various application problems.

Chapter 6

Sup Controllers and Self-Tuning Sup Regulators

6.1 Introduction

In order to control systems with unknown parameters the solution via the method of self-tuning control has received considerable attention. Following the initial work of Astrom and Wittenmark [8], many self-tuning control methods have been studied, such as the minimum variance self-tuner [8], the generalized minimum-variance approach [26, 27], the pole-placement algorithm [9, 162, 163] and the generalized predictive control approach [28]. In these approaches, the system disturbances are assumed to have some probabilistic and/or statistical properties, e.g., white noise with zero mean and finite variance. However, in practice, disturbances seldom meet these assumptions, e.g., disturbances are bounded in both magnitude [35, 124, 148] and increment [91, 92] (for example, step changes of material quality at random times). To cope with systems with disturbances characterized by a bound on their values, some robust adaptive controllers have been developed [81, 119, 121, 133].

This chapter is mainly concerned with the problem of self-tuning control for linear systems with constant but unknown parameters. The external inputs to which the system may be subjected are characterized as a set of inputs bounded on their increments. The chapter begins with the design problem of systems with constant and known parameters, and concludes with the design problem for systems with constant but unknown parameters.

In the design of a controller for these systems one should consider the following important three aspects of:

- Input space: this is concerned with environmental conditions to which the system is subjected to, including reference inputs and disturbances.

- Output performance: this is the ability of the system to achieve satisfactory outputs.

- Stability: this is concerned with stability of the system, including boundness of outputs.

In this chapter, a controller which minimizes the largest absolute value of the generalized output (error or output) of a system for all external inputs which the system is subjected to and for all time, called the sup controller, is developed for systems with the input space $\mathcal{F}(m, \delta)$ or the complex input space $\mathcal{F}(N, m_0, \delta_0)$ (see section 6.2), which contains persistent inputs and/or transient inputs, and with a minimum phase plant. If there is no reference input, the sup controller is called the sup regulator since the design of the system is a regulation problem. Some relations are found between the input space, the sup controller and the optimal output performance. For non-minimum phase plants, the design of sup controllers can be solved via numerical methods based on linear programming and the extrema of functions.

In the case of unknown parameters, since the external input space is not modelled as a white noise with mean zero and finite variance, the general least squares identification algorithm is not suitable to estimate the parameters. A modified least squares algorithm (MLS) is used to estimate the unknown parameters. The self-tuning sup regulator algorithm [99] is obtained on the basis of the MLS estimator and the sup regulator.

The main result is a characterization of the desired closed-loop system when the self-tuning sup regulator algorithm is applied to the system. It is shown that the properties of the self-tuning sup regulator will converge to those of the sup regulator under the assumptions that the system to be controlled is stably invertible, the time delay is known, the upper bound of the order of the system is given and the upper bound on the increment of disturbances is known.

External input spaces, which contain persistent and/or transient inputs, are introduced in Section 6.2. A performance index considering an external input space to which a system is subjected is presented in Section 6.3. In Section 6.4, a sup regulator for systems without reference input is discussed, and it will be shown that the optimal output performance is related to the external input space (here called the disturbance space) and the sup regulator. In Section 6.5, the sup controller is developed and some relations between the input space, the sup controller and the optimal output performance are also given in simple formulae. Sup controllers for systems with a non-minimum phase plant are discussed in Section 6.6. The MLS algorithm and the self-tuning sup regulator algorithm are presented in Section 6.7. The convergence of the closed-loop self-tuning system is analyzed in Section 6.8. Finally, a simulation example is included to demonstrate the applicability of the self-tuning sup regulator in Section 6.9.

6.2   External Input Spaces

The definition of an external input space (or called an input space) plays an important role in control systems since it forms an integral part of the control system representation.

The choice of the external input space is crucial to the definition of performance and directly affects the design method of a system. Two kinds of external input spaces in the discrete-time domain are presented in this section. These input spaces are $\mathcal{F}(m, \delta)$ and $\mathcal{F}(N, m_0, \delta_0)$ for discrete systems [98, 91, 92]. They are defined on the basis of the increment of the inputs rather than the magnitude of the inputs.

Definition 6.1 *The input space $\mathcal{F}(m, \delta)$ is a set of sequences w such that*

$$\sup\left\{\sum_{i=k}^{k+m} |\Delta w(i)| : k \in \mathbb{N}\right\} \leq \delta, \tag{6.2.1}$$

$$|w(k)| < \infty, \quad \text{for} \quad k \in \mathbb{N}, \tag{6.2.2}$$

where $\delta \in (0, \infty)$, $m \in \mathbb{N}^+$ and as usual, Δw is the increment of the external disturbance w, i.e.

$$\Delta w(k) = w(k) - w(k-1). \tag{6.2.3}$$

From Eq.(6.2.1), clearly, the input space $\mathcal{F}(m, \delta)$ essentially implies a restriction on the rate of change of inputs. For $m = 0$, an important feature of the space $\mathcal{F}(0, \delta)$ is that it contains elements w that change substantially for all time $t \in \mathbb{N}^+$ within the constraint

$$\sup\{|\Delta w(k)| : k \in \mathbb{N}\} \leq \delta. \tag{6.2.4}$$

Control systems are frequently subjected to persistent external inputs which can be modelled by elements of $\mathcal{F}(0, \delta)$. Therefore, the external input space $\mathcal{F}(0, \delta)$ is called a bounded increment persistent input space [103]. In contrast, for $m = \infty$, $\mathcal{F}(\infty, \delta)$ contains the elements w which do not change substantially for all time in $\mathbb{N}^+$, but there exist some w such that

$$\lim_{k \to \infty} \Delta w(k) = 0. \tag{6.2.5}$$

$\mathcal{F}(\infty, \delta)$ is known as a transient input space [103].

Therefore, the input space $\mathcal{F}(m, \delta)$ covers persistent and transient inputs. Some properties of the input space $\mathcal{F}(m, \delta)$ are given by the following propositions.

Proposition 6.1 *If $0 < m < \infty$ and $0 < r \leq m$, then*

$$\mathcal{F}(m - r, \frac{\delta}{r+1}) \subset \mathcal{F}(m, \delta). \tag{6.2.6}$$

Proof: For $0 < r \leq m$, there exists the following inequality:

$$\sum_{i=k}^{k+m} |\Delta w(i)| \leq \sum_{i=k}^{k+m-r} |\Delta w(i)| + \sum_{i=k}^{k+m-r} |\Delta w(i+1)|$$

$$+ \ldots + \sum_{i=k}^{k+m-r} |\Delta w(i+r)|. \tag{6.2.7}$$

For all $w \in \mathcal{F}(m - r, \delta(r + 1)^{-1})$, then we have

$$\sum_{i=k}^{k+m-r} |\Delta w(i)| \leq \frac{\delta}{r+1}, \quad \forall k \in \mathbb{N}. \tag{6.2.8}$$

Substituting Eq.(6.2.8) into Eq.(6.2.7) gives

$$\sum_{i=k}^{t+m} |\Delta w(i)| \leq (r+1)\frac{\delta}{r+1} = \delta, \quad \forall t \in \mathbb{N}. \tag{6.2.9}$$

Hence, all $w \in \mathcal{F}(m - r, \delta(r + 1)^{-1})$ belong to $\mathcal{F}(m, \delta)$. i.e.,

$$\mathcal{F}(m - r, \frac{\delta}{r+1}) \subseteq \mathcal{F}(m, \delta). \tag{6.2.10}$$

Also, if $0 < m < \infty$, then

$$\mathcal{F}(m - r, \frac{\delta}{r+1}) \neq \mathcal{F}(m, \delta). \tag{6.2.11}$$

In order to verify this, suppose that $w(k) = 0$ for $k \neq k_0$ and $w(k) = \delta$ for $k = k_0$. Then $w \in \mathcal{F}(m, \delta)$ but $w \notin \mathcal{F}(m - r, \delta(r + 1)^{-1})$.   □□

Proposition 6.2 *If $0 < m < \infty$, then*

$$\mathcal{F}(m, \delta) \subset \mathcal{F}(m - 1, \delta). \tag{6.2.12}$$

The result of Proposition 6.2 is clear from the definition of $\mathcal{F}(m, \delta)$. Propositions 6.1 and 6.2 imply that for $r = 1$,

$$\mathcal{F}(m - 1, \frac{\delta}{2}) \subset \mathcal{F}(m, \delta) \subset \mathcal{F}(m - 1, \delta). \tag{6.2.13}$$

Thus, it shows how larger $\mathcal{F}(m - 1, \delta)$ is compared to $\mathcal{F}(m, \delta)$.

In many practical systems, either the persistent input space or the transient input space cannot effectively describe the external inputs to which the systems are subjected, because an input w is generally made up of several parts, e.g. a persistent part and a transient part. Thus, suppose the input w is the sum of the sequences $w^{(j)}$ ($j = 1, 2, ..., N$), i.e.

$$w = \sum_{j=1}^{N} w^{(j)}, \tag{6.2.14}$$

where each $w^{(j)}$ belongs to a class $\mathcal{F}(m_{0j}, \delta_{0j})$ as defined above. Let

$$m_0 = [m_{01}, m_{02}, ..., m_{0N}], \tag{6.2.15}$$

$$\delta_0 = [\delta_{01}, \delta_{02}, ..., \delta_{0N}]. \tag{6.2.16}$$

Definition 6.2 *The complex input space is defined as*

$$\mathcal{F}(N, m_0, \delta_0) = \left\{ \sum_{j=1}^{N} w^{(j)} : w^{(j)} \in \mathcal{F}(m_{0j}, \delta_{0j}), \quad \text{for} \quad j = 1, 2, ..., N \right\}. \tag{6.2.17}$$

Clearly, the complex input space $\mathcal{F}(N, m_0, \delta_0)$ covers at least a persistent input if one of the elements in m_0 is infinity and at least a transient input if one of elements in m_0 is finite.

6.3 Performance Index

A problem that occurs frequently in control engineering is to control an output v (usually the error) of a system subjected to random external inputs w (reference inputs and/or disturbances) so that the absolute value of the output is within a prescribed bound ε, that is,

$$|v(t, w)| \le \varepsilon, \quad \forall t \in \mathbb{R}. \tag{6.3.1}$$

Any violation of the bound results in unacceptable, and perhaps catastrophic operation. For example, automatically guided motor vehicles are required to travel within their prescribed lanes and any departure from those can have serious consequences. Another example is the control of force on yarns of a loom in textile mills. The force on the yarn is required to remain within specified narrow bounds so that the yarn is not broken, despite disturbances such as quality changes of the yarns at random times. Violation of this requirement results in the loom stopping. Further examples are papermaking and steel press. Such systems are said to be critical [90, 94, 97, 102, 167, 182]. It is well known that the measure of performance in control system design is quite important since its minimisation prescribes the structure and type of controller. A performance function for critical systems is thus defined, based on corresponding input spaces.

Let $\mathcal{F}$ represent all input spaces in the discrete-time domain and the external input vector $\bar{w} = [r, w]$, where r is the reference input and w is the disturbance. The output of the system

$$v(\bar{w}, C) : k \mapsto v(k, \bar{w}, C), \quad k \in \mathbb{N}, \tag{6.3.2}$$

depends upon the external input $\bar{w}$ and the controller C.

Following Zakian [179, 180, 181], the definition of the performance measure for a system is associated with its corresponding input space. Thus, the measure of performance of the system with external inputs in $\mathcal{F}$ is defined as

$$\phi(C) = \sup\{|v(k, \bar{w}, C)| : k \in \mathbb{N}, \bar{w} \in \mathcal{F}\}. \tag{6.3.3}$$

Roughly speaking, $\phi(C)$ is the largest absolute value of the generalized output v for all external inputs $\bar{w}$ in $\mathcal{F}$ as time k ranges over $\mathbb{N}$.

In most control systems, the main objective is to ensure that over all time k, the generalized output remains as small as possible. A control system is said to be well behaved if it optimizes the performance index that includes, among other conditions, a mathematical statement of how the performance function has to be minimized. An appropriate mathematical statement for discrete systems is expressed in the following form:

$$\min_C \phi(C). \tag{6.3.4}$$

The design objective Eq.(6.3.4) gives rise to the problem of designing the controller C that minimizes the performance index. This is difficult to achieve if the evaluation of the number $\phi(C)$ is effected directly from the defining expression Eq.(6.3.4), this is due to the supremum operation being taken over the input function space. To overcome this difficulty, some explicitly simple expressions for $\phi(C)$ are presented in the following sections.

Now, let us give two definitions about polynomials in discrete time which are used in this chapter and next chapter.

Definition 6.3 *Let* $\mathbb{R}[q,p]$ *denote the set of polynomials in the indeterminate* q^{-1} *with coefficients in the field* $\mathbb{R}$ *of real numbers and with the order* p *in* $\mathbb{N}^+$ *of non-negative integer numbers.*

For example, the polynomial $H \in \mathbb{R}[q,n]$, that is, $H = h_0 + h_1 q^{-1} + ... + h_n q^{-n}$.

Definition 6.4 *Let* $\bar{A}_p$ *($p = 1, 2, ..., \infty$) denote the set of polynomials* H *in* $\mathbb{R}[q,\infty]$ *which satisfy*

$$\|H\|_p := \left[\sum_{i=0}^{\infty} |h_i|^p\right]^{1/p} < \infty, \quad p \in [1,\infty), \tag{6.3.5}$$

$$\|H\|_\infty := \sup\{|h_i| : i \in \mathbb{N}^+\} < \infty. \tag{6.3.6}$$

For example, if the polynomial $G \in \bar{A}_1$, then $\|G\| = \sum_{i=0}^{\infty} |g_i| < \infty.$

6.4   Sup Regulators and Output Performance

This section considers a system without the reference input r. In this instance, the generalized output v is replaced by the output y, and the performance function Eq.(6.3.3) for a system is replaced by

$$\phi_{DR}(C) = \sup\{|y(k,\omega,C)| : k \in \mathbb{N}, \omega \in \mathcal{F}\}, \tag{6.4.1}$$

where the input space $\mathcal{F}$ represents either $\mathcal{F}(m,\delta)$ or $\mathcal{F}(N, m_0, \delta_0)$. This will then lead us to consider it as a regulation problem of control systems. The purpose of design is to find a regulator $C : y \mapsto u$ such that the performance function $\phi_{DR}(C)$ is minimized. Such a regulator is called a sup regulator.

For the sake of simplicity, let the ARMAX model describing a plant be of the form

$$B_d y(k) = q^{-d} B_n u(k) + w(k), \qquad (6.4.2)$$

where the polynomials $B_d \in \mathbb{R}[q, \bar{n}]$ with $b_{d0} = 1$ and $B_n \in \mathbb{R}[q, \bar{m}]$, the time delay $d \in \mathbb{N}^+ \setminus 0$, y, u and w are the output, the control and the disturbance of the system, respectively.

Since the polynomials ΔB_d and q^d are coprime, there exists the following identity [28]:

$$\Delta B_d F + q^{-d} E = 1, \qquad (6.4.3)$$

where the polynomials $F \in \mathbb{R}[q, d-1]$ with $f_0 = 1$ and $E \in \mathbb{R}[q, \bar{n}]$ are unique.

Lemma 6.1 *The plant Eq.(6.4.2) can be expressed in a predictor form as*

$$y(k + d) = \xi(k) + \Delta F w(k + d), \qquad (6.4.4)$$
$$\xi(k) = Ey(k) + \Delta F B_n u(k), \qquad (6.4.5)$$

where the polynomials F and E are determined by the identity Eq.(6.4.3).

Proof: Multiplying $\Delta F q^d$ in Eq.(6.4.2) yields

$$\Delta B_d F y(k + d) = \Delta F B_n u(k) + \Delta F w(k + d). \qquad (6.4.6)$$

From the identity Eq.(6.4.3), the lemma follows. $\square\square$

Remark 6.1 *Lemma 6.1 shows that $y(k + d)$ consists of two terms: one depending on the past known controls together with measured outputs and the other depending on the disturbances. As F is of order $d - 1$, the disturbance components are all in the future so that it is impossible to know the value of $\Delta F w(k + d)$ from given measured output y and the control u up to time k.*

Now, let us consider the input space $\mathcal{F} = \mathcal{F}(m, \delta)$. Since different values of m in the input space $\mathcal{F}(m, \delta)$ might lead to different output performance, the following two cases are studied.

Case 1: $0 \leq m < d - 1$
  Let $|\Delta w(i)| = \gamma_i \delta$ for $i \in \mathbb{N}$, where $\gamma_i \geq 0$ $(i \in \mathbb{N})$. For $w \in \mathcal{F}(m, \delta)$, then

$$\sum_{i=k}^{k+m} \gamma_i \leq 1, \quad \forall k \in \mathbb{N}. \qquad (6.4.7)$$

By Lemma 6.1, we have

$$|y(k+d)| \leq |\xi(k)| + \chi(\gamma_{k+1}, \gamma_{k+2}, \ldots, \gamma_{k+d})\delta, \tag{6.4.8}$$

where

$$\chi(\gamma_{k+1}, \gamma_{k+2}, \ldots, \gamma_{k+d}) = \sum_{i=0}^{d-1} |f_i|\gamma_{k+d-i}. \tag{6.4.9}$$

From Eq.(6.4.8), the problem is to find a vector $[\gamma_{k+1}, \gamma_{k+2}, \ldots, \gamma_{k+d}]$ which maximizes χ such that it maximizes the linear form (i.e. the objective function)

$$\sum_{i=0}^{d-1} |f_i|\gamma_{k+d-i} \tag{6.4.10}$$

subject to the linear constraints

$$\gamma_{k+i} \geq 0, \quad i = 1, 2 \ldots, d \tag{6.4.11}$$

and, from Eq.(6.4.7),

$$\sum_{i=j}^{j+m} \gamma_i \leq 1, \quad j = k+1, k+2, \ldots, k+d-m. \tag{6.4.12}$$

In fact, the above problem is a special case of the general linear programming problem. Using the Simplex method [25, 33, 108], the optimal solution $[\gamma_{k+1}^o, \gamma_{k+2}^o, \ldots, \gamma_{k+d}^o]$ can be obtained in a finite number of Simplex stages. Hence

$$\max\{\chi(\gamma_{k+1}, \gamma_{k+2}, \ldots, \gamma_{k+d})\} = \sum_{i=0}^{d-1} |f_i|\gamma_{k+d-i}^o. \tag{6.4.13}$$

It can be seen that this linear programming problem does not depend on a specific $k \in \mathbb{N}$. Therefore, all optimal solutions $[\gamma_{k+1}^o, \gamma_{k+2}^o, \ldots, \gamma_{k+d}^o]$ for $k \in \mathbb{N}$ are same. Let $[\gamma_1^o, \gamma_2^o, \ldots, \gamma_d^o]$ denote such an optimal solution, it can be obtained that

$$\max\{\chi(\gamma_{k+1}, \gamma_{k+2}, \ldots, \gamma_{k+d}) : k \in \mathbb{N}\} = \sum_{i=0}^{d-1} |f_i|\gamma_{d-i}^o. \tag{6.4.14}$$

From Eq.(6.4.8), it can be seen that

$$|y(k+d)| \leq |\xi(k)| + \sum_{i=0}^{d-1} |f_i|\gamma_{d-i}^o\delta \tag{6.4.15}$$

which implies, by Eq.(6.4.1), that

$$\phi_{DR}(C) \leq \sup\{|\xi(k)| : k \in \mathbb{N}\} + \sum_{i=0}^{d-1} |f_i|\gamma_{d-i}^o\delta. \tag{6.4.16}$$

In $\mathcal{F}(m, \delta)$, let us choose a specific disturbance w^* which is defined by

$$w^*(k) = \begin{cases} 0, & k \leq 0 \\ w^*(k-1) + \delta\gamma_k^o \, \mathrm{sgn} \, (f_{d-k}), & 0 < k \leq d \\ w^*(k-1), & k > d \end{cases} \quad (6.4.17)$$

and consider a regulator C_R of the form

$$\Delta F B_n u(k) = -Ey(k). \quad (6.4.18)$$

Since this implies that $\xi(k) = 0$, it follows from Eq.(6.4.4) that

$$y(d, w^*, C_R) = \sum_{i=0}^{d-1} |f_i| \gamma_{d-i}^o \delta. \quad (6.4.19)$$

With Remark 6.1, it is clear from Eqs.(6.4.16) and (6.4.19) that $\phi_{DR}(C)$ can not be smaller than $y(d, w^*, C_R)$. Therefore, if a regulator C can be found such that

$$\phi_{DR}(C) = \sum_{i=0}^{d-1} |f_i| \gamma_{d-i}^o \delta, \quad (6.4.20)$$

then the regulator minimizes $\phi_{DR}(C)$. In fact, the regulator C_R defined by Eq.(6.4.18) together with Eqs.(6.4.16) and (6.4.19) implies that the minimum value of the performance function is

$$\phi_{DR}(C_R) = \sum_{i=0}^{d-1} |f_i| \gamma_{d-i}^o \delta \quad (6.4.21)$$

and the output is

$$y(k, w, C_R) = \Delta F w(k). \quad (6.4.22)$$

To summarize, the following theorem can be readily obtained.

Theorem 6.1 *Consider a system with the plant Eq.(6.4.2) and the input space $\mathcal{F}(m, \delta)$. Then for $0 \leq m < d-1$ in $\mathcal{F}(m, \delta)$, the sup regulator C_R defined by Eq.(6.4.18) minimizes $\phi_{DR}(C)$, and gives the optimal performance Eq.(6.4.21) and the output Eq.(6.4.22).*

Example 6.1 Let us consider the following plant

$$y(k) + 0.7y(k-1) = 0.5u(k-3) + w(k) \quad (6.4.23)$$

with the disturbance $w \in \mathcal{F}(1, 0.1)$.
By solving the identity Eq.(6.4.3), two polynomials, F and E, are obtained

$$F = 1 + 0.3q^{-1} + 0.79q^{-2}, \quad (6.4.24)$$
$$E = 0.447 + 0.553q^{-1}. \quad (6.4.25)$$

From Eq.(6.4.18), the sup regulator C_R is

$$u(k) = -0.894y(k) - 1.106y(k-1) + 0.7u(k-1) - 0.49u(k-2) + 0.79u(k-3). \quad (6.4.26)$$

In this case, the objective function to be minimized for the linear programming is $\gamma_1 + 0.3\gamma_2 + 0.79\gamma_3$ subject to the constraints $\gamma_i \geq 0$, for $i = 1, 2, 3$, $\gamma_1 + \gamma_2 \leq 1$ and $\gamma_2 + \gamma_3 \leq 1$. It can be seen that the optimal solution of the linear programming is $[\gamma_1^o, \gamma_2^o, \gamma_3^o] = [1, 0, 1]$. Therefore,

$$\phi_{DR}(C_R) = (1 + 0.79) \times 0.1 = 0.179. \quad (6.4.27)$$

Remark 6.2 *For the special case $m = 0$, it is apparent that the optimal solution $[\gamma_1^o, \gamma_2^o, \ldots , \gamma_d^o]$ is $[1, 1, ..., 1]$. In this case, the linear programming problem above can simply be stated as the objective function*

$$\sum_{i=0}^{d-1} |f_i| \gamma_{d-i}^o \quad (6.4.28)$$

subject to the linear constraints

$$0 \leq \gamma_i \leq 1, \quad i = 1, 2..., d. \quad (6.4.29)$$

Obviously, the optimal solution of the problem is $\gamma_i = 1$ for $i = 1, 2, ..., d$. Therefore, for $m = 0$,

$$\phi_{DR}(C_R) = \sum_{i=0}^{d-1} |f_i| \delta. \quad (6.4.30)$$

In Example 6.1, if the disturbance $\omega \in \mathcal{F}(0, 0.1)$, then the optimal output performance becomes

$$\phi_{DR}(C_R) = (1 + 0.3 + 0.79) \times 0.1 = 0.209. \quad (6.4.31)$$

Case 2: $m \geq d-1$

In this case, the following theorem can be obtained.

Theorem 6.2 *If $m \geq d-1$ in $\mathcal{F}(m, \delta)$, then the sup regulator C_R that minimizes $\phi_{DR}(C)$ is also given by Eq.(6.4.18) and*

$$\phi_{DR}(C_R) = \|F\|_\infty \delta \quad (6.4.32)$$

In this case, Eq.(6.4.22) still holds.

Proof: It follows from Eq.(6.4.4) that

$$|y(k+d)| \leq |\xi(k)| + |f_0||\Delta\omega(k+d)| + |f_1||\Delta\omega(k+d-1)| + \ldots + |f_{d-1}||\Delta\omega(k+1)|$$
$$\leq |\xi(k)| + \|F\|_\infty \sum_{i=k+1}^{k+d} |\Delta\omega(i)|. \quad (6.4.33)$$

where f_i $(i = 0, 1, ..., d-1)$ is the coefficient of the polynomial F. Since $w \in \mathcal{F}(m, \delta)$ and $m \geq d - 1$, we have

$$|y(k + d)| \leq |\xi(k)| + \|F\|_\infty \delta. \tag{6.4.34}$$

By using Eq.(6.4.1), it can be obtained that

$$\phi_{DR}(C) \leq \sup\{|\xi(k)| : k \in \mathbb{N}\} + \|F\|_\infty \delta. \tag{6.4.35}$$

Similar to Eq. (6.4.17), let us choose another particular disturbance

$$\Delta w^*(k) = \begin{cases} \delta \ \text{sgn} \ (f_M), & k = M \\ 0, & k \neq M \end{cases} \tag{6.4.36}$$

where f_M belongs to $\{f_0, f_1, \ldots, f_{d-1}\}$ and has the greatest absolute value in $\{f_0, f_1, \ldots, f_{d-1}\}$. Clearly, $w^* \in \mathcal{F}(m, \delta)$. Consider the regulator C_R defined by Eq.(6.4.18). Since this implies that $\xi(k) = 0$, it follows that

$$y(d, w^*, C_R) = \delta |f_M| = \|F\|_\infty \delta. \tag{6.4.37}$$

With the same reasoning as that for Case 1, the theorem follows from Eq.(6.4.35) and Eq.(6.4.37) immediately. $\quad\square\square$

In Example 6.1, assume the disturbance $w \in \mathcal{F}(2, 0.1)$. Then the minimum of the performance function is $\phi_{DR}(C_R) = 0.1$.

From these two cases, it is clear that the sup regulator C_R is independent of the parameters m and δ of the input space $\mathcal{F}(m, \delta)$.

Define

$$Q_R(m) = \begin{cases} \sum_{i=0}^{d-1} |f_i| \gamma^o_{d-i}, & 0 \leq m < d - 1 \\ \|F\|_\infty, & m \geq d - 1 \end{cases} \tag{6.4.38}$$

where γ^o_k $(k = 1, 2, .., d)$ is the same as that in Case 1.

A generalization of Theorems 6.1 and 6.2 is obtained for systems with the input space $\mathcal{F} = \mathcal{F}(N, m_0, \delta_0)$ as follows.

Theorem 6.3 *Suppose that the disturbance $w \in \mathcal{F}(N, m_0, \delta_0)$. Then the sup regulator C_R defined by Eq.(6.4.18) minimizes $\phi_{DR}(C)$ and leads to*

$$\phi_{DR}(C_R) = \sum_{i=1}^{N} Q_R(m_{0i})\delta_{0i}, \tag{6.4.39}$$

$$y(k, w, C_R) = F \sum_{i=1}^{N} \Delta w^{(i)}(k), \tag{6.4.40}$$

where the polynomial F is the same as that in Eq.(6.4.3).

Proof: Immediate from the combination of Theorems 6.1 and 6.2. $\square\square$

To illustrate theorem 6.3, assume that the disturbance $w \in \mathcal{F}(3, [0, 1, 2], [0.02, 0.03, 0.05])$ in Example 6.1. Then, by Theorem 6.3, $\phi_{DR}(C_R) = 2.09 \times 0.02 + 1.79 \times 0.03 + 1 \times 0.05 = 0.1455$.

Summarizing the results of Theorems 6.1-6.3 results in

$$\phi_{DR}(C_R) = \begin{cases} Q_R(m)\delta, & \mathcal{F} = \mathcal{F}(m, \delta) \\ \sum_{i=1}^{N} Q_R(m_{0i})\delta_{0i}, & \mathcal{F} = \mathcal{F}(N, m_0, \delta_0) \end{cases} \qquad (6.4.41)$$

which provides a formula for computing the optimal performance function and displays some simple relations between the input space $\mathcal{F}$, the sup regulator C_R and the optimal performance $\phi_{DR}(C_R)$.

Having established the general structure of the sup regulators, the stability of the closed loop system needs to be evaluated. Since the reference input $r = 0$, the plant Eq.(6.4.2) and the sup regulator Eq.(6.4.18) lead to the following closed-loop system equation:

$$\begin{aligned} y(k) &= \frac{\Delta F B_n B_d^{-1}}{\Delta B_n F + E B_n q^{-d} B_d^{-1}} w(k) \\ &= \frac{\Delta B_n F}{\Delta F B_d B_n + E B_n q^{-d}} w(k). \end{aligned} \qquad (6.4.42)$$

Using the identity Eq.(6.4.3) gives

$$y(k) = \frac{\Delta F B_n}{B_n} w(k). \qquad (6.4.43)$$

Obviously, the characteristic equation of the closed-loop system is

$$B_n = 0. \qquad (6.4.44)$$

If the zeros of the polynomial B_n do not lie within the closed unit disc of the complex plane, then the system is stable and the term B_n in Eq.(6.4.43) can be cancelled without effecting stability. It can then be concluded that the results in this section are only useful for minimum phase plants. The design of a system with a non-minimum phase plant will be discussed in Section 6.6.

6.5   Sup Controllers and Output Performance

This section is concerned with the problem of controlling the error $e = r - y$ of a system with $\bar{w} \in \mathcal{F}_1 \times \mathcal{F}_2$. In this case, the reference input r ranges over $\mathcal{F}_1$, the disturbance w

ranges over $\mathcal{F}_2$, where $\mathcal{F}_1$ denotes either $\mathcal{F}(m^{(1)}, \delta^{(1)})$ or $\mathcal{F}(N, m_0^{(1)}, \delta_0^{(1)})$ and in a similar way $\mathcal{F}_2$ denotes either $\mathcal{F}(m^{(2)}, \delta^{(2)})$ or $\mathcal{F}(N, m_0^{(2)}, \delta_0^{(2)})$. Then, following Eq.(6.3.3), the performance function is defined as

$$\phi_D(C) = \sup\{|e(k, \bar{w}, C)| : k \in \mathbb{N}, \bar{w} \in \mathcal{F}_1 \times \mathcal{F}_2\}, \tag{6.5.1}$$

where the external input vector $\bar{w} = [r, w]$. Let the plant description of a system be the same as Eq.(6.4.2), i.e.,

$$B_d y(k) = q^{-d} B_n u(k) + w(k). \tag{6.5.2}$$

To simplify the presentation we first consider a special case in which the disturbance $w = 0$, so that the plant Eq.(6.5.2) becomes the ARMA model:

$$B_d y(k) = q^{-d} B_n u(k). \tag{6.5.3}$$

The performance function Eq.(6.5.1) is simplified to read

$$\phi_{DC}(C) = \sup\{|e(k, r, C)| : k \in \mathbb{N}, r \in \mathcal{F}_1\}. \tag{6.5.4}$$

Define

$$Q_C(m) = \begin{cases} \sum_{i=0}^{d-1} \gamma_i^o, & 0 \leq m < d - 1 \\ 1, & m \geq d - 1 \end{cases} \tag{6.5.5}$$

where γ_i^o ($i = 0, 1, ..., d - 1$) is the same as that of Section 6.4, the following theorem can be obtained.

Theorem 6.4 *For system Eq.(6.5.3) and the reference input $r \in \mathcal{F}_1$, the sup controller C_O, which minimizes $\phi_{DC}(C)$, is given by*

$$\Delta F B_n u(k) = r(k) - E y(k), \tag{6.5.6}$$

and the minimum value of the performance function is

$$\phi_{DC}(C_O) = \begin{cases} Q_C(m^{(1)})\delta^{(1)}, & \mathcal{F}_1 = \mathcal{F}(m^{(1)}, \delta^{(1)}) \\ \sum_{i=1}^{N} Q_C(m_{0i}^{(1)})\delta_{0i}^{(1)}, & \mathcal{F}_1 = \mathcal{F}(N^{(1)}, m_0^{(1)}, \delta_0^{(1)}) \end{cases} \tag{6.5.7}$$

and the error

$$e(k, r, C_O) = r(k) - r(k - d). \tag{6.5.8}$$

Proof: Since the disturbance $w = 0$, it follows from Lemma 6.1 that

$$y(k + d) = \xi(k). \tag{6.5.9}$$

Subtracting $r(k + d)$ from both sides of Eq.(6.5.9) gives

$$
\begin{aligned}
y(k + d) - r(k + d) &= \xi(k) - r(k + d) \\
&= \xi(k) - r(k) - (r(k + d) - r(k)). \quad\quad (6.5.10)
\end{aligned}
$$

Since $e(k) = r(k) - y(k)$, we have

$$
e(k + d) = r(k) - \xi(k) + (r(k + d) - r(k)). \quad\quad (6.5.11)
$$

Let $r(k) - \xi(k)$ and $r(k + d) - r(k)$ replace $\xi(k)$ and $\Delta Fw(k)$ in Cases 1 and 2 of Section 6.4, respectively. Then using the same reasoning as Cases 1 and 2, the results of the theorem follow immediately.   □□

Note that the sup controller Eq.(6.5.6) is directly obtained by setting $\xi(k) = r(k)$. We now consider the general case in which there are both the reference input r and the disturbance w.

Theorem 6.5 *Suppose that $\bar{w} \in \mathcal{F}_1 \times \mathcal{F}_2$ in a system. Then the sup controller C_O given by Eq.(6.5.6) minimizes $\phi_D(C)$ and leads to*

$$
\phi_D(C_O) = \phi_{DC}(C_O)|_{\mathcal{F}_1} + \phi_{DR}(C_R)|_{\mathcal{F}_2} \quad\quad (6.5.12)
$$

and the minimal error is

$$
e(k, \bar{w}, C_O) = \sum_{i=1}^{d} \Delta q^{-i}(r(k) - f_{d-i}w(k)), \quad\quad (6.5.13)
$$

where $(.)|_{\mathcal{F}_j}$ $(j = 1, 2)$ means the condition under which the input space is $\mathcal{F}_j$, and $\phi_{DC}(C_O)$ and $\phi_{DR}(C_R)$ are given by Eq.(6.5.7) and Eq.(6.4.41), respectively.

Proof: Using Lemma 6.1 and the fact that $e(k) = r(k) - y(k)$, it can be seen that

$$
e(k, \bar{w}, C) = r(k) - \xi(k) + (r(k + d) - r(k)) - \Delta Fw(k + d). \quad\quad (6.5.14)
$$

From the previous results, it follows that

$$
\begin{aligned}
\sup\{|r(k + d) - r(k)| : k \in \mathbb{N}, r \in \mathcal{F}_1\} &\leq \phi_{DC}(C_O)|_{\mathcal{F}_1}, \quad\quad (6.5.15) \\
\sup\{|\Delta Fw(k + d)| : k \in \mathbb{N}, w \in \mathcal{F}_2\} &\leq \phi_{DR}(C_R)|_{\mathcal{F}_2}. \quad\quad (6.5.16)
\end{aligned}
$$

which give, with Eqs.(6.5.1) and (6.5.14),

$$
\phi_D(C_O) \leq \sup\{|r(k) - \xi(k)| : k \in \mathbb{N}\} + \phi_{DC}(C_O)|_{\mathcal{F}_1} + \phi_{DR}(C_R)|_{\mathcal{F}_2}. \quad\quad (6.5.17)
$$

According to the manner of choosing the particular disturbance in Section 6.4, a particular reference input r^* in $\mathcal{F}_1$ and a particular disturbance w^* in $\mathcal{F}_2$ can be found such that

$$r^*(k+d) - r^*(k) = \phi_{DC}(C_O)|_{\mathcal{F}_1}, \tag{6.5.18}$$

$$-\Delta F w^*(k+d) = \phi_{DR}(C_R)|_{\mathcal{F}_2}. \tag{6.5.19}$$

Consider the sup controller C_O given by Eq.(6.5.6). Since it implies $\xi(k) = r(k)$, it follows from Eqs.(6.5.18) and (6.5.19) that

$$e(k, \bar{w}^*, C_O) = \phi_{DC}(C_O)|_{\mathcal{F}_1} + \phi_{DR}(C_R)|_{\mathcal{F}_2}. \tag{6.5.20}$$

With Eqs.(6.5.17) and (6.5.20), it can be obtained that

$$\phi_D(C_O) = \phi_{DC}(C_O)|_{\mathcal{F}_1} + \phi_{DR}(C_R)|_{\mathcal{F}_2}. \tag{6.5.21}$$

Substituting $\xi(k) = r(k)$ into Eq.(6.5.14) gives

$$e(k, \bar{w}, C_O) = r(k) - r(k-d) - \Delta F w(k) \tag{6.5.22}$$

$$= \sum_{i=1}^{d} \Delta q^{-i}(r(k) - f_{d-i} w(k)). \tag{6.5.23}$$

$\square\square$

The following example illustrates the properties of the sup controller.

Example 6.2 Let the plant be given by

$$y(k) + 0.5y(k-1) + 0.1y(k-2) = u(k-3) + 0.8u(k-4) + w(k). \tag{6.5.24}$$

with the reference input r and the disturbance w being chosen as

$$r(k+1) = r(k) + \begin{cases} 0.03, & k \in [0, 200) \\ 0 + 0.02 \ \text{sgn} \ (\alpha(k)), & k \in [200, 800) \\ -0.03, & k \in [800, \infty) \end{cases} \tag{6.5.25}$$

$$w(k+1) = w(k) + 0.08 \ \text{sgn} \ (\beta(k)), \tag{6.5.26}$$

where $\alpha(k)$ and $\beta(k)$ are random variables ranged over $[-1, 1]$ with zero mean and with variance equal to one. Clearly, the external input vector $[r, w] \in \mathcal{F}(0, 0.05) \times \mathcal{F}(0, 0.08)$. Using Theorem 6.4, the sup controller for this plant is described by

$$\begin{aligned} u(k) = \ & r(k) - 0.625y(k) - 0.31y(k-1) - 0.065y(k-2) - 0.3u(k-1) + \\ & +0.25u(k-2) + 0.53u(k-3) + 0.52u(k-4). \end{aligned} \tag{6.5.27}$$

and

$$F(q^{-1}) = 1 + 0.5q^{-1} + 0.65q^{-2}. \tag{6.5.28}$$

As a result, $\phi_{DC}(C_O) = 0.15$, $\phi_{DR}(C_R) = 0.172$ and $\phi_D(C_O) = 0.322$. In other words, the sup controller limits the error of the system exactly to the expected bounds $\pm\phi_D(C_O)$.

To analyze the stability of the closed-loop system with the sup controller, the following closed-loop system equation is formulated from the plant Eq.(6.5.2) and the sup controller Eq.(6.5.6)

$$
\begin{aligned}
e(k) &= \frac{-\Delta F B_n B_d^{-1}}{\Delta B_n F + E B_n q^{-d} B_d^{-1}} w(k) - \frac{B_n q^{-d} B_d^{-1} - \Delta B_n F - E B_n q^{-d} B_d^{-1}}{\Delta B_n F + E B_n q^{-d} B_d^{-1}} r(k) \\
&= \frac{-\Delta B_n F}{\Delta F B_d B_n + E B_n q^{-d}} w(k) - \frac{B_n q^{-d} - \Delta F B_d B_n - E B_n q^{-d}}{\Delta F B_d B_n + E B_n q^{-d}} r(k).
\end{aligned} \tag{6.5.29}
$$

Using the identity Eq.(6.4.3), it can be seen that

$$
e(k) = \frac{B_n}{B_n}((1 - q^{-d}) r(k) - \Delta F w(k)). \tag{6.5.30}
$$

Clearly, the characteristic equation of the closed-loop system is

$$
B_n = 0. \tag{6.5.31}
$$

Again, if the zeros of the polynomial B_n are not within the closed unit disc of the complex plane, then the system is stable and the term B_n Eq.(6.5.30) can be cancelled without effecting stability. Therefore the results in this section are only for minimum phase plants.

6.6 Sup Controllers for Non-Minimum Phase Plants

In this section, we present the design procedure of sup controllers for a non-minimum phase plant with input space $\mathcal{F}_1 \times \mathcal{F}_2$, where $\mathcal{F}_1$ and $\mathcal{F}_2$ are as the same as those in Section 6.5. The performance function is

$$
\phi_D(C) = \sup\{|e(k, \bar{w}, C)| : k \in \mathbb{N}, \bar{w} \in \mathcal{F}_1 \times \mathcal{F}_2\}, \tag{6.6.1}
$$

where the external input vector $\bar{w} = [r, w]$; the error $e = r - y$; y, r and w are the output, the reference input and the disturbance of the system, respectively.

As before, consider a plant having the ARMAX model

$$
B_d y(k) = q^{-d} B_n u(k) + w(k), \tag{6.6.2}
$$

where $B_d \in \mathbb{R}[q, \bar{n}]$ and $B_n \in \mathbb{R}[q, \bar{m}]$.

The following assumptions about the plant are made: i) the polynomials ΔB_d and $q^{-d} B_n$ are coprime. ii) B_n has some zeros within the closed unit disc.

From part (i) of the assumption, the following identity [28]

$$\Delta B_d Q + q^{-d} B_n P = 1, \tag{6.6.3}$$

will hold, where the polynomials $Q \in \mathbb{R}[q, d + \bar{m} - 1]$, with $q_0 = 1$ and $P \in \mathbb{R}[q, \bar{n}]$, are unique.

Using the YJB parameterization of all stabilizing controllers [175], the sup controller has the following form

$$(\Delta Q - \bar{R}\Delta B_n q^{-d})u(k) = (P + \bar{R}\Delta B_d)e(k), \quad \bar{R} \in \bar{A}_1, \tag{6.6.4}$$

where P and Q are given by Eq.(6.6.3).

From Eq.(6.6.2) and Eq.(6.6.4), it can be seen that the error e is related to the reference input r and the disturbance w in the following way

$$e(k) = (Q - \bar{R}B_n q^{-d})B_d \Delta r(k) - (Q - \bar{R}B_n q^{-d})\Delta w(k). \tag{6.6.5}$$

Assume that B_n is factorized as a product BB_s, where B has all zeros inside the closed unit disc whilst B_s has all zeros outside the closed unit disc. Let $R = \bar{R}B_s$, where $\bar{R} \in \bar{A}_1$, then it can be seen that $R \in \bar{A}_1$ because B_s has all stable zeros. Therefore, $Q - \bar{R}B_n q^{-d} = Q - RBq^{-d}$. Consequently, the sup controller Eq.(6.6.4) becomes

$$(\Delta Q B_s - \Delta R B_n q^{-d})u(k) = (PB_s + \Delta R B_d)e(k), \quad R \in \bar{A}_1, \tag{6.6.6}$$

and the relation between e, r and w is

$$e(k) = (Q - RBq^{-d})B_d \Delta r(k) - (Q - RBq^{-d})\Delta w(k). \tag{6.6.7}$$

It is clear that the sup controller depends on the unknown polynomial $R \in \bar{A}_1$. Thus, we replace $\phi_D(C)$ by $\phi_D(R)$. The design problem is to find an $R \in \bar{A}_1$ such that $\phi_D(R)$ is minimized. Let $H^c = (Q - RBq^{-d})B_d \in \mathbb{R}[q, N_c]$ and $H^r = (Q - RBq^{-d}) \in \mathbb{R}[q, N_r]$, where $N_c \in [0, \infty]$ and $N_r \in [0, \infty]$. For the sake of simplicity, we assume that $\mathcal{F}_1 = \mathcal{F}(m_1, \delta_1)$, $\mathcal{F}_2 = \mathcal{F}(m_2, \delta_2)$ and $\min\{N_c, N_r\} > \max\{m_1, m_2\}$. For $k \in \mathbb{N}$, let

$$|\Delta r(k - i)| = a_i \delta_1, \quad i = 0, 1, ..., N_c, \tag{6.6.8}$$
$$|\Delta w(k - j)| = a_j \delta_2, \quad j = 0, 1, ..., N_r. \tag{6.6.9}$$

Since $r \in \mathcal{F}(m_1, \delta_1)$ and $w \in \mathcal{F}(m_2, \delta_2)$, we have

$$a_i \geq 0, \quad i = 0, 1, ..., N_c, \tag{6.6.10}$$
$$\sum_{l=i}^{i+m_1} a_l \leq 1, \quad i = 0, 1, ..., N_c - m_1, \tag{6.6.11}$$
$$b_j \geq 0, \quad j = 0, 1, ..., N_r, \tag{6.6.12}$$
$$\sum_{l=j}^{j+m_2} b_l \leq 1, \quad j = 0, 1, ..., N_r - m_2. \tag{6.6.13}$$

Using the same reasoning as that in Case 1 in Section 6.4, it follows that the performance function can be expressed by

$$\phi_D(R) = \max_{\mathbf{a}} \chi_c(R, \mathbf{a})\delta_1 + \max_{\mathbf{b}} \chi_r(R, \mathbf{b})\delta_2 \qquad (6.6.14)$$

where

$$\chi_c(R, \mathbf{a}) = \sum_{l=0}^{N_c} |h_l^c| a_l, \qquad (6.6.15)$$

$$\chi_c(R, \mathbf{b}) = \sum_{l=0}^{N_r} |h_l^r| b_l, \qquad (6.6.16)$$

$$\mathbf{a} = [a_0, a_1, \dots, a_{N_c}], \qquad (6.6.17)$$
$$\mathbf{b} = [b_0, b_1, \dots, b_{N_r}], \qquad (6.6.18)$$

and h_k^c and h_k^r are the coefficients of the polynomials H^c and H^r, respectively.

Therefore, the sup controller is given by

$$\min_{R \in \bar{A}_1} \{\max_{\mathbf{a}} \chi_c(R, \mathbf{a})\delta_1 + \max_{\mathbf{b}} \chi_r(R, \mathbf{b})\delta_2\}, \qquad (6.6.19)$$

subject to Eqs.(6.6.10)-(6.6.13).

Although it is very difficult to obtain an analytical solution for the above problem, it can be solved by numerical methods. Obviously, this is a minimax problem which consists of two loops: the internal loop (the max part) and the external loop (the min part). The internal loop,

$$\max_{\mathbf{a}} \chi_c(R, \mathbf{a})\delta_1 + \max_{\mathbf{b}} \chi_r(R, \mathbf{b})\delta_2 \qquad (6.6.20)$$

subject to Eqs.(6.6.10)-(6.6.13), is a linear programming problem, which can be solved by numerous algorithms, such as, the Simplex algorithm [25, 33]. The external loop, $\min \phi_D(R)$ subject to $R \in \bar{A}_1$, is a problem of minimizing a function subject to a constraint.

Remark 6.3 *By setting the reference input $r = 0$, a special case, a regulation problem, is obtained. In this case, the structure of sup regulators is same as that of the sup controllers Eq.(6.6.7), but the performance function is simpler than that Eq.(6.6.14) and becomes*

$$\max_{\mathbf{b}} \chi_r(R, \mathbf{b})\delta_2, \qquad (6.6.21)$$

which is subject to Eq.(6.6.12) and Eq.(6.6.13). The sup regulator can be obtained by carrying out

$$\min_{R \in \bar{A}_1} \max_{\mathbf{b}} \chi_r(R, \mathbf{b}) \qquad (6.6.22)$$

which can also be solved by certain numerical methods.

Even though it has not been proved whether there exists a bound for the order of the polynomial R such that $\phi_D(R)$ is minimized, in many cases when the order of R increases to a finite number that leads to a significant reduction of $\phi_D(R)$, little improvement of $\phi_D(R)$ results from further increase of the order of R. We do not intend to discuss the design of sup controllers for the complex input spaces and $\min\{N_c, N_r\} \le \max\{m_1, m_2\}$, but the above methods could be similarly applied.

6.7 The Self-Tuning Sup Regulator Algorithm

Consider a plant described by

$$B_d y(k) = q^{-d} B_n u(k) + w(k), \tag{6.7.1}$$

where the polynomials $B_d \in \mathbb{R}[q, \bar{n}]$ with $b_{d0} = 1$ and $B_n \in \mathbb{R}[q, \bar{m}]$, the time delay $d \in \mathbb{N}^+ \backslash 0$, y, u and w are the output, the control and the disturbance of the system, respectively.

To simplify the following discussion, we assume that the disturbance $w \in \mathcal{F}(0, \delta)$, which means that the disturbances are bounded on their increments by δ. For the plant described by Eq.(6.7.1), if the parameters of the plant are known, the sup regulator of the form Eq.(6.4.18), i.e.

$$\Delta F B_n u(k) = -E y(k), \tag{6.7.2}$$

can be constructed, where polynomials $F \in \mathbb{R}[q, d - 1]$ and $E \in \mathbb{R}[q, \bar{n}]$ are given by Eq.(6.4.3).

However, if the parameters are unknown, it is natural that the estimated parameters should be used to form the controller. The only difference is that the true parameters of the plant is substituted by their estimates. In fact, this is the basic idea of self-tuning control. A modified least squares (MLS) algorithm is described in this section to estimate the parameters of the plant Eq.(6.7.1) with a bound on the increment of disturbances.

To simplify our notation, we denote y_k, u_k and w_k for $y(k)$, $u(k)$ and $w(k)$, respectively, and moreover

$$G = 1 - \Delta B_d, \tag{6.7.3}$$

where

$$G = g_1 q^{-1} + g_2 q^{-2} + \dots + g_{\bar{n}+1} q^{-\bar{n}-1}. \tag{6.7.4}$$

Then the plant Eq.(6.7.1) can be expressed as

$$y_k = \theta^T \psi_{k-1} + \Delta w_k, \tag{6.7.5}$$

where

$$\theta = [g_1, g_2, \dots, g_{\bar{n}+1}, b_{n0}, b_{n1}, \dots, b_{n\bar{m}}]^T, \tag{6.7.6}$$

$$\psi_{k-1} = [y_{k-1}, y_{k-2}, ..., y_{k-\bar{n}-1}, \Delta u_{k-d}, \Delta u_{k-d-1}, ..., \Delta u_{k-d-\bar{m}}]^T, \qquad (6.7.7)$$

$$\Delta u_k = u_k - u_{k-1}, \qquad (6.7.8)$$

$$\Delta \omega_k = \omega_k - \omega_{k-1}. \qquad (6.7.9)$$

Note that later in this chapter we use $\hat{x}$ to denote the estimate of x. The estimation problem is then to find a θ belonging to the set defined by

$$\Xi(\theta) = \{\theta : |y_k - \theta^T \psi_{k-1}| \le \delta, \forall k \in \mathbb{N}\}. \qquad (6.7.10)$$

Three different types of solutions to this problem have mainly been explored in literature. The first method is to formulate the estimation problem in a geometrical setting [50]. The second approach is to derive estimation algorithm from stability consideration together with the geometrical approach [122]. The third alternative is obtain estimation algorithm from modifying the exponentially weighted recursive least squares algorithm [21]. However, the following limitations have been observed.

i) Fogel and Huang's algorithm [50] is more complicated than the other two algorithms so that more computational work is needed.

ii) In Ortega and Lozano-Leal's algorithm [122], the parameter estimation is carried out as

$$\hat{\theta}_k = \hat{\theta}_{k-1} + \frac{\alpha P_{k-1}\psi_{k-1}}{1 + \psi_{k-1}^T P_{k-1}\psi_{k-1}}(|e_k| - \delta_1)\ \text{sgn}\ (e_k), \qquad \alpha \in (0,1), \qquad (6.7.11)$$

$$P_k^{-1} = \begin{cases} P_{k-1}^{-1} + \dfrac{\alpha P_{k-1}\psi_{k-1}}{(1 + \psi_{k-1}^T P_{k-1}\psi_{k-1})e_k}(|e_k| - \delta_1)\ \text{sgn}\ (e_k), & |e_k| > \delta_1 \\ P_{k-1}^{-1}, & |e_k| \le \delta_1 \end{cases} \qquad (6.7.12)$$

$$\delta_1 = \sqrt{1 + \alpha}\delta, \qquad (6.7.13)$$

$$e_k = y_k - \hat{\theta}_{k-1}^T \psi_{k-1}. \qquad (6.7.14)$$

If it is assumed that $\psi_{k-1}^T P_{k-1}\psi_{k-1}$ is finite, then this algorithm shows that

$$\lim_{k \to \infty} |e_k| = \sqrt{1 + \alpha}\delta, \qquad \alpha \in (0,1). \qquad (6.7.15)$$

This means that the prediction output error e_k can not be less than or equal to the bound δ.

iii) As for Canudas and Carrillo's algorithm [21], we have

$$\hat{\theta}_k = \hat{\theta}_{k-1} + \frac{\alpha_k P_{k-1}\psi_{k-1}}{\psi_{k-1}^T P_{k-1}\psi_{k-1}}(|e_k| - \delta)\ \text{sgn}\ (e_k), \qquad (6.7.16)$$

$$P_k = \lambda^{-1}\left(P_{k-1} - \frac{\alpha_k P_{k-1}\psi_{k-1}\psi_{k-1}^T P_{k-1}}{\psi_{k-1}^T P_{k-1}\psi_{k-1}}(1 - \frac{\delta}{|e_k|})\right), \qquad \lambda \in (0,1], \qquad (6.7.17)$$

$$\alpha_k = \begin{cases} 1, & |e_k| > \delta \quad \text{or} \quad \psi_{k-1}^T P_{k-1} \psi_{k-1} = 0 \\ 0, & |e_k| \le \delta \end{cases} \tag{6.7.18}$$

$$e_k = y_k - \hat{\theta}_{k-1}^T \psi_{k-1}. \tag{6.7.19}$$

In this algorithm, when $|e_k| > \delta$ and $\psi_{k-1}^T P_{k-1} \psi_{k-1}$ is very small rather than zero, the parameter estimate $\hat{\theta}_k$ might not converge because the term

$$P_{k-1} \psi_{k-1} (\psi_{k-1}^T P_{k-1} \psi_{k-1})^{-1} \tag{6.7.20}$$

may be quite large. On the other hand, if α_k is set to zero, then $|e_k| \le \delta$ can not be guaranteed when $\psi_{k-1}^T P_{k-1} \psi_{k-1}$ is finite.

In order to remove these limitations, the following new modified least squares identification algorithm for the plant Eq.(6.7.1) is formulated.

Theorem 6.6 *Consider the plant Eq.(6.7.1) and the identification algorithm:*

$$\hat{\theta}_k = \hat{\theta}_{k-1} + \frac{\alpha_k P_{k-1} \psi_{k-1}}{1 + \psi_{k-1}^T P_{k-1} \psi_{k-1}} (|e_k| - \delta) \; \text{sgn} \; (e_k), \tag{6.7.21}$$

$$P_k = P_{k-1} - \frac{\alpha_k P_{k-1} \psi_{k-1} \psi_{k-1}^T P_{k-1}}{|e_k| + (2|e_k| - \delta) \psi_{k-1}^T P_{k-1} \psi_{k-1}} (|e_k| - \delta), \tag{6.7.22}$$

$$e_k = y_k - \hat{\theta}_{k-1}^T \psi_{k-1}, \tag{6.7.23}$$

$$\alpha_k = \begin{cases} 1, & |e_k| > \delta \\ 0, & |e_k| \le \delta \end{cases} \tag{6.7.24}$$

Then

$$i) \qquad \lim_{k \to \infty} \frac{\alpha_k (|e_k| - \delta)^2}{1 + \psi_{k-1}^T P_{k-1} \psi_{k-1}} = 0, \tag{6.7.25}$$

$$ii) \qquad \lim_{k \to \infty} |\hat{\theta}_k - \hat{\theta}_{k-1}| = 0, \tag{6.7.26}$$

$$iii) \qquad \left\| \tilde{\theta}_k \right\|^2 \le \frac{\lambda_{max}(P_0^{-1})}{\lambda_{min}(P_0^{-1})} \left\| \tilde{\theta}_0 \right\|^2, \tag{6.7.27}$$

where

$$\tilde{\theta}_k = \theta - \hat{\theta}_k \tag{6.7.28}$$

and $\lambda_{max}(.)$ and $\lambda_{min}(.)$ denote the maximum and the minimum eigenvalues of the matrix (.).

Proof:

i) Let

$$V_k = \tilde{\theta}_k^T P_k^{-1} \tilde{\theta}_k \qquad (6.7.29)$$

be a Lyapunov function, then by combining Eqs.(6.7.5), (6.7.21)-(6.7.23), it can be obtained that

$$V_k = V_{k-1} + \frac{\alpha_k(|e_k| - \delta)}{(1 + \psi_{k-1}^T P_{k-1} \psi_{k-1})|e_k|} \left[\Delta\omega_k^2 - \frac{(1 + \psi_{k-1}^T P_{k-1} \psi_{k-1})|e_k|^3}{|e_k| + (2|e_k| - \delta)\psi_{k-1}^T P_{k-1} \psi_{k-1}} \right]. \qquad (6.7.30)$$

For $\omega \in \mathcal{F}(0, \delta)$, it can be seen that the following inequality

$$V_k \leq V_{k-1} + \frac{\alpha_k(|e_k| - \delta)}{(1 + \psi_{k-1}^T P_{k-1} \psi_{k-1})|e_k|} \left[\delta^2 - \frac{(1 + \psi_{k-1}^T P_{k-1} \psi_{k-1})|e_k|^3}{|e_k| + (2|e_k| - \delta)\psi_{k-1}^T P_{k-1} \psi_{k-1}} \right]$$

$$= V_{k-1} - \frac{\alpha_k(|e_k| - \delta)}{(1 + \psi_{k-1}^T P_{k-1} \psi_{k-1})|e_k|} \times$$

$$\times \frac{(|e_k|^3 - 2|e_k|\delta^2 + \delta^3)(1 + \psi_{k-1}^T P_{k-1} \psi_{k-1}) + (|e_k| - \delta)\delta^2}{|e_k| + (2|e_k| - \delta)\psi_{k-1}^T P_{k-1} \psi_{k-1}}$$

$$\leq V_{k-1} - \frac{\alpha_k(|e_k| - \delta)(|e_k|^3 - 2|e_k|\delta^2 + \delta^3)}{(|e_k| + (2|e_k| - \delta)\psi_{k-1}^T P_{k-1} \psi_{k-1})|e_k|} \qquad (6.7.31)$$

holds since $|e_k| \geq \delta$. In addition, for $|e_k| \geq \delta$,

$$e_k^2 - 2\delta^2 + \frac{\delta^3}{|e_k|} \geq |e_k|(|e_k| - \delta). \qquad (6.7.32)$$

Therefore

$$V_k \leq V_{k-1} - \frac{\alpha_k(|e_k| - \delta)^2}{2(1 + \psi_{k-1}^T P_{k-1} \psi_{k-1})} \qquad (6.7.33)$$

and

$$\lim_{k \to \infty} \frac{\alpha_k(|e_k| - \delta)^2}{1 + \psi_{k-1}^T P_{k-1} \psi_{k-1}} = 0. \qquad (6.7.34)$$

ii) From Eq.(6.7.21), it can be seen that

$$\left\| \hat{\theta}_k - \hat{\theta}_{k-1} \right\|^2 = \frac{\alpha_k \psi_{k-1}^T P_{k-1}^2 \psi_{k-1}}{(1 + \psi_{k-1}^T P_{k-1} \psi_{k-1})^2}(|e_k| - \delta)^2$$

$$\leq \frac{\alpha_k \lambda_{max}(P_{k-1})}{1 + \psi_{k-1}^T P_{k-1} \psi_{k-1}}(|e_k| - \delta)^2. \qquad (6.7.35)$$

Since

$$\lambda_{max}(P_k) \leq \lambda_{max}(P_{k-1}) \leq \ldots \leq \lambda_{max}(P_0) \qquad (6.7.36)$$

Eq.(6.7.35) can be further written as

$$\left\| \hat{\theta}_k - \hat{\theta}_{k-1} \right\|^2 \le \alpha_k \lambda_{max}(P_0) \frac{(|e_k| - \delta)^2}{1 + \psi_{k-1}^T P_{k-1} \psi_{k-1}}, \tag{6.7.37}$$

As a result, part (ii) of the theorem is proved.

iii) Using the well known matrix inversion theorem [57], it follows that

$$P_k^{-1} = P_{k-1}^{-1} + \frac{\alpha_k \psi_{k-1} \psi_{k-1}^T}{1 + \psi_{k-1}^T P_{k-1} \psi_{k-1}} \left(1 - \frac{\delta}{|e_k|} \right). \tag{6.7.38}$$

Therefore

$$\lambda_{min}(P_k^{-1}) \ge \lambda_{min}(P_{k-1}^{-1}) \ge \ldots \ge \lambda_{min}(P_0^{-1}). \tag{6.7.39}$$

Using Eq.(6.7.33), together with the formulations performed, we have

$$V_k \le V_0, \tag{6.7.40}$$

$$\lambda_{min}(P_0^{-1}) \left\| \tilde{\theta}_k \right\|^2 \le \lambda_{max}(P_k^{-1}) \left\| \tilde{\theta}_0 \right\|^2. \tag{6.7.41}$$

As a result, it can be obtained that

$$\left\| \tilde{\theta}_k \right\|^2 \le \frac{\lambda_{max}(P_0^{-1})}{\lambda_{min}(P_0^{-1})} \left\| \tilde{\theta}_0 \right\|^2 \tag{6.7.42}$$

which establishes part (iii) of the theorem. □□

The properties (i)-(iii) are instrumental in establishing stability of the self-tuning sup regulator. In order to control a system with time-invariant unknown parameters and with a bound on the increment of disturbances, a self-tuning sup regulator algorithm is presented on the basis of the sup regulator in Section 6.4 and the MLS algorithm. The following assumptions are made.

Assumption 6.1

i) *The time delay d is known.*

ii) *An upper bound for the orders of the polynomials B_d and B_n in the plant Eq.(6.7.1) is known.*

iii) *The polynomial B_n has all zeros outside the closed unit disc.*

Note that part (iii) of Assumption 6.1 is necessary to achieve closed-loop stability for the sup regulator when the parameters are known, which is given in section 6.4. Thus this is a natural assumption in the case where the parameters are unknown. Part (ii) of Assumption 6.1 is significant, since it allows the order of the true plant not to be overestimated. Part (i) of Assumption 1 is vital and necessary to ensure that $b_{n0} \neq 0$. This is needed in the subsequent formulation. The self-tuning sup regulator algorithm is given as follows:

i) **Step 1: Parameter Estimation**

At the sampling interval k determine the parameter θ of the model

$$y_k = \theta^T \psi_{k-1} + \Delta \omega_k \tag{6.7.43}$$

using the MLS algorithm (based on all data available at time k) and then compute the polynomials F and E from

$$\hat{F}_k(1 - \hat{G}_k) + \hat{E}_k q^{-d} = 1, \tag{6.7.44}$$

where $\hat{F}_k$, $\hat{G}_k$ and $\hat{E}_k$ are the estimated values of the polynomials F, G and E at time k, respectively, and $\hat{f}_0(k) = 1$.

ii) **Step 2: Control**

The objective is to design a control law such that the output satisfies

$$\lim_{k \to \infty} \sup |y_k| \leq \left\| \hat{F} \right\|_1 \delta, \tag{6.7.45}$$

where

$$\hat{F} = \lim_{k \to \infty} \hat{F}_k. \tag{6.7.46}$$

The predictable output is

$$\hat{y}_{k+d} = \hat{E}_k y_k + \hat{F}_k \hat{B}_{nk} \Delta u(k). \tag{6.7.47}$$

Therefore, at each sampling interval determine the control variable from the sup regulator C_R^k:

$$\hat{E}_k y_k + \hat{F}_k \hat{B}_{nk} \Delta u(k) = 0 \tag{6.7.48}$$

where the polynomials $\hat{E}_k$, $\hat{F}_k$ and $\hat{B}_{nk}$ are obtained in Step 1.

Since the MLS algorithm is computed recursively, the algorithm requires only moderate computations.

6.8 Convergence of the Self-Tuning Algorithm

Using the technical lemma described in chapter 2, the following theorem, regarding the global convergence of the self-tuning sup regulator algorithm, can be established.

Theorem 6.7 *Subject to Assumption 6.1, the self-tuning sup regulator algorithm Eq.(6.7.2 1)-Eq.(6.7.24) and Eq.(6.7.48), when applied to the plant Eq.(6.7.1), yields*

$$i) \quad y_k \quad \text{and} \quad u_k \quad \text{are bounded sequences.} \tag{6.8.1}$$

$$ii) \quad \limsup_{k \to \infty} |e_k| \le \delta. \tag{6.8.2}$$

$$iii) \quad \limsup_{k \to \infty} |y_k| \le \left\| \hat{F} \right\|_1 \delta \tag{6.8.3}$$

Proof: It is known from Eq.(6.7.23) that

$$e_k = y_k - \hat{\theta}_{k-1}^T \psi_{k-1} \tag{6.8.4}$$

Let

$$\bar{y}_k = \hat{\theta}_{k-d}^T \psi_{k-1} \tag{6.8.5}$$

and define a track error ε_k as

$$\varepsilon_k = y_k - \bar{y}_k \tag{6.8.6}$$

then by pre-multiplying Eq.(6.8.5) with $\hat{F}_{k-d}$, it can be obtained that

$$
\begin{aligned}
\hat{F}_{k-d}\bar{y}_k &= \hat{F}_{k-d}\hat{\theta}_{k-d}^T \psi_{k-1} \\
&= \hat{F}_{k-d}\hat{G}_{k-d}y_k + \hat{F}_{k-d}q^{-d}(\hat{B}_{nk}\Delta u(k)) \\
&= \hat{F}_{k-d}(\hat{G}_{k-d} - 1)y_k + \hat{F}_{k-d}y_k + \hat{F}_{k-d}q^{-d}(\hat{B}_{nk}\Delta u(k)) \\
&= (\hat{F}_{k-d} - 1)y_k + \hat{E}_{k-d}y_{k-d} + \hat{F}_{k-d}q^{-d}(\hat{B}_{nk}\Delta u(k)), \quad \text{using} \quad \text{Eq.(6.7.44)} \\
&= (\hat{F}_{k-d} - 1)y_k + q^{-d}(\hat{E}_k y_k + \hat{F}_k \hat{B}_{nk}\Delta u(k)) \\
&= (\hat{F}_{k-d} - 1)y_k, \quad \text{using} \quad \text{Eq.(6.7.48).} \tag{6.8.7}
\end{aligned}
$$

Therefore

$$y_k = \hat{F}_{k-d}(y_k - \bar{y}_k) = \hat{F}_{k-d}\varepsilon_k \tag{6.8.8}$$

and

$$|y_k| \le m_1 \max_{\tau \le k} |\varepsilon_\tau|, \tag{6.8.9}$$

with

$$m_1 = \max_{k \in \mathbb{N}} \left\| \hat{F} \right\|_1 \tag{6.8.10}$$

From part (iii) of Theorem 6.6, it can be shown that m_1 is finite.

Since the plant Eq.(6.7.1) is assumed to be a linear discrete-time stably invertible system [57], we have

$$|u(k-d)| \le m_2 + m_3 \max_{\tau \le k} |y(\tau)|, \quad 0 \le m_2, m_3 < \infty \tag{6.8.11}$$

As a result

$$\| \psi_{k-1} \| \le (\bar{m}+1)m_2 + (\bar{n}+\bar{m}+2)\max\{1, m_3\} \max_{\tau \le k} |y(\tau)| \tag{6.8.12}$$

Combining Eqs.(6.8.9) and (6.8.12) gives

$$\| \psi_{k-1} \| \le (\bar{m}+1)m_2 + (\bar{n}+\bar{m}+2)\max\{1, m_3\}m_1 \max_{\tau \le k} |\varepsilon_\tau|,$$

$$\le C_1 + C_2 \max_{\tau \le k} |\varepsilon_\tau|, \quad 0 \le C_1, C_2 < \infty. \tag{6.8.13}$$

From Eqs.(6.8.4)-(6.8.6), it can be shown that

$$\varepsilon_k = e_k - (\tilde{\theta}_{k-1}^T - \tilde{\theta}_{k-d}^T)\psi_{k-1}. \tag{6.8.14}$$

Thus

$$\frac{\alpha_k|(|\varepsilon_k| - \delta)|}{(1 + \psi_{k-1}^T P_{k-1}\psi_{k-1})^{1/2}} \le \frac{\alpha_k|(|e_k| - \delta)| + \alpha_k|(\tilde{\theta}_{k-1}^T - \tilde{\theta}_{k-d}^T)\psi_{k-1}|}{(1 + \psi_{k-1}^T P_{k-1}\psi_{k-1})^{1/2}}. \tag{6.8.15}$$

From parts (i) and (ii) of Theorem 6.6, it can be seen that the right-hand side of Eq. (6.8.15) will tend to zero when $k \to \infty$. Hence

$$\lim_{k \to \infty} \frac{\alpha_k(|\varepsilon_k| - \delta)^2}{1 + \psi_{k-1}^T P_{k-1}\psi_{k-1}} = 0 \tag{6.8.16}$$

Using Eq.(6.7.36), we have

$$\frac{\alpha_k(|\varepsilon_k| - \delta)^2}{1 + \psi_{k-1}^T P_{k-1}\psi_{k-1}} \ge \frac{\alpha_k(|\varepsilon_k| - \delta)^2}{1 + \lambda_{max}(P_0)\| \psi_{k-1} \|2}. \tag{6.8.17}$$

If $\alpha_k = 0$ as $k \to \infty$, then Eqs.(6.8.1) and (6.8.2) are proved. Otherwise, $\alpha_k \ne 0$ as $k \to \infty$. Thus let

$$\sigma_1 = 1, \quad \sigma_2 = \lambda_{max}(P_0) \quad \text{and} \quad \phi(k) = \psi_{k-1}. \tag{6.8.18}$$

Using Eq.(2.7.1) of Lemma 2.1, Eqs.(6.8.1) and (6.8.2) can still be obtained. From part (ii) of Theorem 6.6, we finally have

$$\lim_{k \to \infty} \hat{F}_{k-d} = \lim_{k \to \infty} \hat{F}_k = \hat{F}. \tag{6.8.19}$$

Since $|\varepsilon_k| \to \delta$ as $k \to \infty$, it follows from Eq.(6.8.8) that

$$\lim_{k \to \infty} \sup |y_k| \leq \left\| \hat{F} \right\|_1 \delta. \tag{6.8.20}$$

Therefore, the theorem is proved. □□

The key conclusions are that

i) Closed-loop stability is achieved.

ii) The output of the system is asymptotically limited to the bounds $\pm \left\| \hat{F} \right\|_1 \delta$.

6.9 An Example

The performance of the self-tuning sup regulator scheme is demonstrated through the following simulation example.

Example 6.3 Consider a plant

$$y(k) + 0.5y(k-1) + 0.1y(k-2) = u(k-3) + 0.8u(k-4) + w(k), \tag{6.9.1}$$

where the disturbance is chosen as

$$w(k) = w(k-1) + 0.08 \ \text{sgn} \ (\nu(k)) \tag{6.9.2}$$

and $\nu(k)$ is a random variable with $E\{\nu(k)\} = 0$ and $E\{\nu^2(k)\} = 1$. Clearly, $w \in \mathcal{F}(0, 0.08)$.

The sup regulator C_R for the system is therefore given by

$$u(k) = -0.625y(k) - 0.31y(k-1) - 0.065y(k-2) - 0.3u(k-1)$$

$$+0.25u(k-2) + 0.53u(k-3) + 0.52u(k-4). \tag{6.9.3}$$

Using Eq.(6.4.3), the polynomial $F = 1 + 0.5q^{-1} + 0.65q^{-2}$. Hence, the optimal output performance

$$\phi_{DR}(C_R) = 0.172. \tag{6.9.4}$$

The self-tuning sup regulator with the structure Eq.(6.7.48) is obtained by using the algorithm described in Section 6.7. After the parameters converged, $\hat{F} = 1 + 0.5004q^{-1} + 0.6503q^{-2}$. As a result

$$\lim_{k \to \infty} \sup \phi_{DR}(C_R^k) \leq 0.172056. \tag{6.9.5}$$

It can be seen from the simulation results that the output of the system is limited to the bound ± 0.172056 after the parameters converge. Figure 6.1 shows the output $y(k)$ of the system. The parameter error norm $\left\| \theta - \hat{\theta}_k \right\|^2$ is displayed in Fig. 6.2.

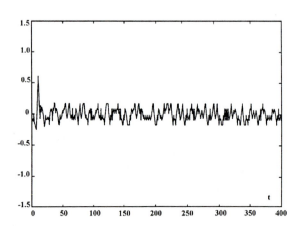

Figure 6.1: The output of the self-tuning system.

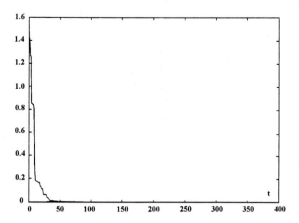

Figure 6.2: The parameter error norm $\left\| \theta - \hat{\theta}_k \right\|^2$.

5.10 Conclusions

This chapter is mainly concerned with the control problem of critical systems with either known or unknown parameters. In the design procedure three important aspects are considered, they are input space, output performance and stability. The input space describes the environmental conditions to which the system is subjected, including the reference inputs and the disturbances, whilst the output performance reflects the ability of the system to reach satisfactory responses.

The input space plays an important role in the design of control systems since it forms an integral part of the closed loop representation, which is an improvement over conventional control theory that only considers typical test inputs, such as step, pulse, sine, white noise and etc. Both the reference input and the disturbance of a system are referred to as the external input. Two kinds of input spaces, $\mathcal{F}(m, \delta)$ and $\mathcal{F}(N, m_0, \delta_0)$, are described. Those input spaces essentially imply a restriction on the rates of changes of inputs and cover transient and persistent inputs.

The measure of performance for control system design is an important issue. In this chapter, a performance function for critical systems is defined as the supremum of the absolute value of the output of the system subjected to all external inputs, which is related to corresponding input spaces.

For systems with input spaces $\mathcal{F}(m, \delta)$ and $\mathcal{F}(N, m_0, \delta_0)$, a controller which minimizes the above performance function is called the sup controller. Particularly, for systems with only disturbance the sup controller is referred to as the sup regulator. A number of relationships between input space, output performance and controller are given for both regulation and control problems. The stability of closed-loop systems with sup regulators and sup controllers have been established.

In order to control the systems with disturbances $\mathcal{F}(0, \delta)$ and with constant unknown parameters, a self-tuning sup regulator is introduced and the sup regulator with the modified least squares algorithm is therefore obtained. The convergence analysis of the closed-loop system shows that the properties of the self-tuning sup regulator asymptotically converge to those of the sup regulator.

Chapter 7

Mean Controllers and Self-Tuning Mean Regulators

7.1 Introduction

In this chapter we turn our attention to the control of linear time invariant stochastic systems and consider both random reference inputs and random disturbances. The process disturbance in stochastic systems is usually assumed to be a zero mean stationary white noise process with a finite variance. Based on this assumption, many methods of designing stochastic control systems have been developed. Examples are the Kalman filtering approach [76], the minimum variance control method [7], the Linear Quadratic Gaussian (LQG) method [132, 171] etc. However, in some control applications the available prior information about the process disturbance is given, not as a zero mean stationary white noise process with a finite variance, but in other forms, such as the bound on its value and the bound on the mean of its increment. Therefore, it is an interesting problem as how to design stochastic systems with other forms of external inputs.

This chapter describes the design procedure of stochastic systems subjected to random input space characterized by a bound on the mean of the absolute increment of the external inputs. The inputs considered are the reference input and the disturbance. A controller called mean controller in literature [100], is discussed. This type of controller minimizes the supremum of the mean of the absolute value of the generalized output for all external inputs characterized by a space, defined in Section 7.2, and for all time.

This chapter is also concerned with the design of self-tuning control for systems with constant unknown parameters. The self-tuning mean regulator algorithm is obtained on the basis of the MLS estimator for the stochastic systems and the mean regulator.

The main result is a characterization of the closed-loop system obtained when the self-tuning mean regulator algorithm is applied to the system. It is shown that the properties of the self-tuning mean regulator will converge to those of the mean regulator under the assumptions that, (i) the system to be controlled is stably invertible, (ii) the time delay s known, (iii) the upper bound of the order of the system is given and (iv) the upper bound on the mean of the increment of disturbances is known.

In Section 7.3. the analysis, which is restricted to zero reference input, yields a special mean controller. This controller is called the mean regulator [100]. The mean controller for systems subject to both reference inputs and disturbances is then derived in Section 7.4, whilst in Section 7.5 the stability of the closed-loop systems is discussed. The self-tuning mean regulator algorithm is finally presented in Section 7.6.

7.2 Input Spaces and Performance Index

Even though for many control systems it is difficult to know external inputs accurately, they may be prescribed by some simple probability distribution. In stochastic control theory the external input is usually assumed to be a zero mean stationary white noise process with finite variance [7, 57], i.e.,

$$E\{w(k)\} = 0, \tag{7.2.1}$$

$$E\{w^2(k)\} < \infty, \tag{7.2.2}$$

where w is the external input of the system and $E\{.\}$ stands for the mean function of a stochastic sequence. However, in practice there are few inputs satisfying these conditions, therefore in this section we use the mean of the input increment to describe an input space.

Definition 7.1 *The input space* $\mathbb{B}(\delta)$ *is defined as a set of mean functions of the absolute value of any stochastic sequence* Δw *such that*

$$\sup\{E\{|\Delta w(k)|\} : k \in \mathbb{N}\} \leq \delta, \tag{7.2.3}$$

$$|w(k)| < \infty, \quad \text{for} \quad k \in \mathbb{N}, \tag{7.2.4}$$

where $\delta \in (0, \infty)$ *and* Δw *is the same as Eq.(6.2.3).*

The input space $\mathbb{B}(\delta)$ is characterized by the number δ, which represents a uniform bound on the mean value of the absolute difference Δw. Following Zakian [179, 180, 181], the performance measure for a system is defined with its corresponding input space in this section. Let $\bar{w} = [r, w]$. The output of a system

$$v(\bar{w}, C) : k \mapsto v(k, \bar{w}, C), \quad k \in \mathbb{N}, \tag{7.2.5}$$

depends upon the external input $\bar{w}$ and the controller C. In the stochastic case, the performance of the system with external inputs $\bar{w}$ and $\bar{w}_E \in \mathbb{B}$ is measured by

$$\phi_S(C) = \sup\{E\{|v(k, \bar{w}, C)|\} : k \in \mathbb{N}, \bar{w}_E \in \mathbb{B}\}. \tag{7.2.6}$$

where $\bar{w}_E(k) = E\{|\Delta\bar{w}(k)|\}$.

It can be seen that $\phi_S(C)$ is the largest mean of the absolute value of the generalized output for $\bar{w}_E \in \mathbb{B}$ and $k \in \mathbb{N}$. A main difficulty is encountered in minimizing the performance index. It means that the evaluation of the number $\phi_S(C)$ is difficult i effected directly from the defining expression Eq.(7.2.6). This is mainly because of the supremum operation required over the input function space. To overcome this difficulty some explicitly simple expressions for $\phi_S(C)$ are given in the following sections.

7.3   Mean Regulators and Output Performance

In this section we consider a linear system (7.3.2) subject to the disturbance w with $w_E \in \mathbb{B}(\delta)$ but without the reference input r, and we assume that the generalized output v is the output y of the plant. Thus, following Eq.(7.2.6), the performance function i defined as

$$\phi_{SR}(C) = \sup\{E\{|y(k, w, C)|\} : k \in \mathbb{N}, w_E \in \mathbb{B}(\delta)\}. \tag{7.3.1}$$

In the above $w_E(k) = E\{|\Delta w(k)|\}$. Clearly, $\phi_{SR}(C)$ is the largest mean of the absolut value of the system output y as $w_E \in \mathbb{B}(\delta)$ for all time k in $\mathbb{N}$. The design problem is to find a feedback regulator $C : y \mapsto u$ such that $\phi_{SR}(C)$ is minimized, called th mean regulator. Without loss of generality, we further assume that the plant is described by a stochastic autoregressive moving-average model with auxiliary input (i.e., ARMA model).

$$B_d y(k) = q^{-d} B_n u(k) + D_n w(k). \tag{7.3.2}$$

where $B_d \in \mathbb{R}[q, \bar{n}]$ with $b_{d0} = 1$, $B_n \in \mathbb{R}[q, \bar{m}]$, $D_n \in \mathbb{R}[q, \bar{l}]$ with $D_{n0} = 1$, y, u and w ar the output, the control and the disturbance of the system, respectively, $\mathbb{R}[q, .]$ is define in Definition 6.3.

For the polynomials B_d and D_n, the following identity [28]:

$$\Delta B_d F + q^{-d} E = D_n, \tag{7.3.3}$$

can still hold, where the polynomials $F \in \mathbb{R}[q, d-1]$ and $E \in \mathbb{R}[q, \bar{n}]$ are uniquel determined.

Initial Condition 7.1 For $k \leq k_0$, the external input $\bar{w}$, the control u and the outpu y of the system are all zeros.

To formulate the main result, the following lemmas are needed.

Lemma 7.1 *Suppose a function x satisfying*

$$x(k) = 0, \quad for \quad k \leq k_0, \tag{7.3.4}$$
$$Ax(k) = 0, \quad for \quad k > k_0, \tag{7.3.5}$$

where the polynomial $A \in \mathbb{R}[q, n]$ with $a_0 = 1$. Then

$$x(k) = 0, \quad \forall k \in \mathbb{N}. \tag{7.3.6}$$

Proof: From Eq.(7.3.5), it follows that for $k > k_0$

$$x(k_0 + 1) = -a_1 x(k_0) - \dots - a_n x(k_0 - n),$$

$$x(k_0 + 2) = -a_1 x(k_0 + 1) - \dots - a_n x(k_0 + 1 - n),$$

$$\dots\dots\dots\dots$$

$$x(k_0 + N) = -a_1 x(k_0 + N - 1) - \dots - a_n x(k_0 + N - 1 - n),$$

$$\dots\dots\dots\dots$$

Since $x(k) = 0$ for $k \le k_0$, it is clear from the above that $x(k_0 + i) = 0$, for $i = 1, 2, \dots$. $\square\square$

Lemma 7.2 *For the plant Eq.(7.3.2), there exists a predictor of the form:*

$$D_n y(k + d) = Ey(k) + \Delta F B_n u(k) + D_n F \Delta \omega(k + d), \tag{7.3.7}$$

where F and E are determined by Eq.(7.3.3).

Proof: Premultiplying Eq.(7.3.2) by $\Delta F q^d$ gives

$$\Delta F B_d y(k + d) = \Delta F B_n u(k) + \Delta F D_n \omega(k + d). \tag{7.3.8}$$

Using Eq.(7.3.3), Eq. (7.3.7) can be readily obtained. $\square\square$

Lemma 7.3 *If the plant satisfies Initial Condition 7.1, then the predictor form of the plant*

$$D_n y(k + d) = Ey(k) + \Delta F B_n u(k) + D_n F \Delta \omega(k + d), \tag{7.3.9}$$

is equivalent to the following form

$$\begin{cases} y(k + d) = \zeta(k) + F \Delta \omega(k + d) \\ D_n \zeta(k) = Ey(k) + \Delta F B_n u(k) \end{cases} \tag{7.3.10}$$

where $\zeta(k) = 0$ for $k \le k_0$.

Proof: i) Eq.(7.3.9) $\rightarrow$ Eq.(7.3.10)

   Let $\zeta(k) = 0$ for $k \le k_0$ and

$$Ey(k) + \Delta F B_n u(k) = D_n \zeta(k). \tag{7.3.11}$$

then Eq.(7.3.9) becomes

$$D_n y(k+d) = D_n \zeta(k) + D_n F \Delta w(k+d). \tag{7.3.12}$$

Hence

$$D_n(y(k+d) - \zeta(k) - F \Delta w(k+d)) = 0. \tag{7.3.13}$$

Using Lemma 7.1, Eq.(7.3.10) can be formulated.

ii) Eq.(7.3.10) $\rightarrow$ Eq.(7.3.9)
Premultiplying the first equation of Eq.(7.3.10) by D_n gives

$$D_n y(k+d) = D_n \zeta(k) + D_n F \Delta w(k+d). \tag{7.3.14}$$

Substituting the second equation of Eq.(7.3.10) into Eq. (7.3.14) leads to Eq.(7.3.9).
Therefore, the result of the lemma is established.   □□

In the equivalent predictor equation Eq.(7.3.10), the output $y(k+d)$ consists of two
terms. The first term $\zeta(k)$ can be computed from the second equation of Eq.(7.3.10).
However, the second term $F \Delta w(k+d)$ is unknown at time k since the disturbance w is
not available.

Theorem 7.1 *Suppose that a system with $w_E \in \mathbb{B}(\delta)$ and $r = 0$ satisfies Initial Condition
7.1. Then the mean regulator $\bar{C}_R$ that minimizes $\phi_{SR}(C)$ is given by*

$$\Delta F B_n u(k) = -E y(k), \tag{7.3.15}$$

and gives

$$\phi_{SR}(\bar{C}_R) = \|F\|_1 \delta, \tag{7.3.16}$$

$$y(k, w, \bar{C}_R) = \Delta F w(k), \tag{7.3.17}$$

*where the polynomials F and E are obtained by solving Eq.(7.3.3), $\|.\|_1$ is defined in
Section 6.3.*

Proof: From Lemma 7.3, it is known that the d-step-ahead prediction output is expressed
equivalently by Eq.(7.3.10). For $w_E \in \mathbb{B}(\delta)$, taking absolute and then mean of the first
equation of Eq.(7.3.10) yields

$$E\{|y(k+d)|\} \le E\{|\zeta(k)|\} + \|F\|_1 \delta. \tag{7.3.18}$$

Since $\zeta(k)$ is a deterministic function at time k,

$$E\{|y(k+d)|\} \le |\zeta(k)| + \|F\|_1 \delta. \tag{7.3.19}$$

Clearly, y and ζ are function of k, ω and C and more explicit notations are $y(k, \omega, C)$ and $\zeta(k, \omega, C)$. According to the performance function Eq.(7.3.1), it can be deduced from Eq.(7.3.19) that

$$\phi_{SR}(C) \leq \sup\{|\zeta(k, \omega, C)| : k \in \mathbb{N}\} + \|F\|_1 \, \delta. \tag{7.3.20}$$

Consider the random disturbance ω^* defined by

$$\omega^*(k) = \begin{cases} 0 & k \leq k_1 \\ \omega^*(k-1) + \delta \text{ sgn } (f_{d+k_1-k}\zeta(k_1, \omega^*, C)) & k_1 < k \leq k_1 + d \\ \omega^*(k-1) & k > k_1 + d \end{cases} \tag{7.3.21}$$

where $k_1 \geq k_0$. Clearly, $\omega_E^* \in \mathbb{B}(\delta)$. Substitute Eq.(7.3.21) into the first equation of Eq.(7.3.10), it can be obtained that

$$y(k_1 + d, \omega^*, C) = \zeta(k_1, \omega^*, C) + \delta \, \|F\|_1 \text{ sgn } (\zeta(k_1, \omega^*, C)). \tag{7.3.22}$$

Hence

$$|y(k_1 + d, \omega^*, C)| = |\zeta(k_1, \omega^*, C)| + \delta \, \|F\|_1 \, . \tag{7.3.23}$$

Using Eq.(7.3.1), we have

$$\phi_{SR}(C) \geq \inf\{|\zeta(k, \omega, C)| : k \in \mathbb{N}\} + \delta \, \|F\|_1 \, . \tag{7.3.24}$$

Combining Eq.(7.3.20) and Eq.(7.3.24), the following inequality

$$\inf\{|\zeta(k, \omega, C)| : k \in \mathbb{N}\} \leq \phi_{SR}(C) - \delta \, \|F\|_1 \leq$$
$$\leq \sup\{|\zeta(k, \omega, C)| : k \in \mathbb{N}\} \tag{7.3.25}$$

can be obtained. It can be seen that if $\zeta(k, \omega, C) = 0, \forall k \in \mathbb{N}$, then $\phi_{SR}(C)$ is minimized. Thus, with the initial conditions and by setting $\zeta(k, \omega, C) = 0 \ (\forall k \in \mathbb{N})$, the mean regulator $\bar{C}_R$ (7.3.15) is obtained. The optimal output performance is given by

$$\phi_{SR}(\bar{C}_R) = \|F\|_1 \, \delta. \tag{7.3.26}$$

Since $\zeta(k, \omega, C) = 0, \forall k \in \mathbb{N}$, it follows from Eq.(7.3.10) that Eq.(7.3.17) holds. $\quad\square\square$

In theorem 7.1 we have derived a mean regulator based on the predictor form of the plant. The mean regulator, using past control and output data, cancels the predictable part of the output on the future response. This fact is clearly illustrated in the following example.

Example 7.1 Consider the following system without reference input

$$y(k) + 0.8y(k-1) + 0.18y(k-2) = u(k-3)$$
$$-0.4u(k-4) + \omega(k) + 0.2\omega(k-1). \tag{7.3.27}$$

with random disturbance being generated from

$$\omega(k) = \omega(k-1) + (d_1 + \varepsilon(k)) \text{ sgn } (\mu(k)), \qquad (7.3.28)$$

where

$$d_1 = \begin{cases} 0.08 & k \in [0, 500] \\ 0.05 & k \in (500, \infty) \end{cases} \qquad (7.3.29)$$

μ and ε are random variables ranged over $[0.02, -0.02]$ with zero mean and with variance equal to 0.1. It can be seen that $\omega_E \in \text{IB}(0.08)$. The mean regulator $\bar{C}_R$ for the system is therefore given by

$$
\begin{aligned}
u(k) &= -0.568y(k) - 0.506y(k-1) - 0.126y(k-2) \\
&\quad + u(k-1) - 0.54u(k-2) + 0.82u(k-3) - 0.28u(k-4) \qquad (7.3.30)
\end{aligned}
$$

and

$$\phi_{SR}(\bar{C}_R) = 0.168. \qquad (7.3.31)$$

Figures 7.1 and 7.2 show the behaviour of the system when the mean regulator (7.3.15) is used. This example illustrates that the mean regulator $\bar{C}_R$ can guarantee the mean of the absolute outputs of the system limited to the prescribed bound, which in this case is 0.168.

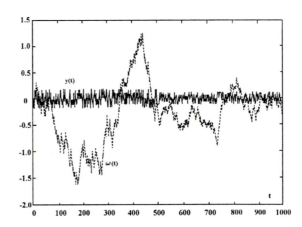

Figure 7.1: The disturbance of the system.

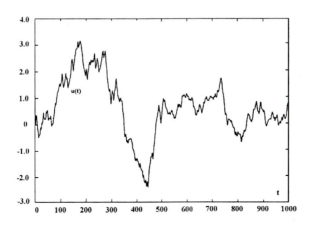

Figure 7.2: The output of the system.

7.4 Mean Controllers and Output Performance

In this section, we consider linear systems subject to both the reference input r and the disturbance w, and we assume that $\bar{w}_E \in \mathbb{B}(\delta_1) \times \mathbb{B}(\delta_2)$ with $\bar{w}_E = [r_E, w_E] = [E\{|\Delta r(k)|\}, E\{|\Delta w(k)|\}]$. The generalized output v is the error e of the system

$$e(k) = r(k) - y(k). \tag{7.4.1}$$

As a result, the performance function (7.3.1) can be rewritten to give

$$\phi_S(C) = \sup\{E\{|e(k, \bar{w}, C)|\} : k \in \mathbb{N}, \bar{w}_E \in \mathbb{B}(\delta_1) \times \mathbb{B}(\delta_2)\}, \tag{7.4.2}$$

where the external input vector $\bar{w} = [r, w]$.

The purpose here is to find a controller $C : (r, y) \mapsto u$ such that $\phi_S(C)$ is minimized, the resultant controller is called the mean controller. Without loss of generality, we still assume that the plant is described by Eq.(7.3.2).

Lemma 7.4 *The error e of the system can be predicted by*

$$D_n e(k + d) = D_n r(k + d) - Ey(k) - \Delta F B_n u(k) - D_n F \Delta w(k + d), \tag{7.4.3}$$

where the polynomials F and E are obtained by solving Eq.(7.3.3).

Proof: It is known from Lemma 7.2 that

$$D_n y(k + d) = Ey(k) + \Delta F B_n u(k) + D_n F \Delta w(k + d). \tag{7.4.4}$$

The lemma can then be proved by subtracting $D_n r(k + d)$ from both sides of Eq. (7.4.4) and by multiplying by -1. $\quad\square\square$

Lemma 7.5 *Suppose that the system considered satisfies the Initial Condition 7.1, then the prediction form Eq.(7.4.3) of the error e is equivalent to the following form*

$$\begin{cases} e(k+d) = (r(k) - \zeta(k)) + \rho(k+d) + F\Delta w(k+d) \\ D_n\zeta(k) = Ey(k) + \Delta F B_n u(k) \\ \rho(k+d) = r(k+d) - r(k) \end{cases} \qquad (7.4.5)$$

where $\zeta(k) = 0$, for $k \leq k_0$.

Proof: Similar to the proof of Lemma 7.3.   □□

In Eq.(7.4.5), the output $e(k+d)$ consists of three terms. The first term $r(k) - \zeta(k)$ can be computed from the input u and output y of the plant up to time k. However, the second term $\rho(k+d)$ (the future reference inputs) and the third term $F\Delta w(k+d)$ (the future disturbances) are unknown because they can not be predicted at time k.

Theorem 7.2 *Suppose that the system with $\bar{w}_E \in \mathbb{B}(\delta_1) \times \mathbb{B}(\delta_2)$ satisfies the Initial Condition 7.1, then the mean controller $\bar{C}_O$ that minimizes $\phi_S(C)$ is given by*

$$\Delta F B_n u(k) = D_n r(k) - Ey(k) \qquad (7.4.6)$$

and

$$\phi_S(\bar{C}_O) = d\delta_1 + \|F\|_1 \delta_2, \qquad (7.4.7)$$
$$e(k, \bar{w}, \bar{C}_O) = \rho(k) - \Delta F w(k), \qquad (7.4.8)$$

where ρ, F and E are the same as those in Lemma 7.5.

Proof: From Lemma 7.5, the d-step-ahead predicted error e of the system can be expressed by Eq.(7.4.5). For $\bar{w}_E \in \mathbb{B}(\delta_1) \times \mathbb{B}(\delta_2)$, taking absolute value and then mean of the first equation of Eq.(7.4.5) gives

$$E\{|e(k+d)|\} \leq E\{|r(k) - \zeta(k)|\} + d\delta_1 + \|F\|_1 \delta_2 \qquad (7.4.9)$$

which can be further written as

$$E\{|e(k+d)|\} \leq |r(k) - \zeta(k)| + d\delta_1 + \|F\|_1 \delta_2 \qquad (7.4.10)$$

as $r(k)$ and $\zeta(k)$ are deterministic functions at time k. It can be seen that e and ζ are the functions of k, $\bar{w}$ and C, therefore they can be denoted explicitly as $e(k, \bar{w}, C)$ and $\zeta(k, \bar{w}, C)$. Using Eq.(7.4.2), it can be deduced from Eq.(7.4.10) that

$$\phi_S(C) \leq \sup\{|r(k) - \zeta(k, \bar{w}, C)| : k \in \mathbb{N}\} + d\delta_1 + \|F\|_1 \delta_2. \qquad (7.4.11)$$

Again, let us choose a particular external input $\bar{w}^* = [r^*, w^*]$ as

$$r^*(k) = \begin{cases} 0 & k \le k_1 \\ r^*(k-1) + \delta_1 \text{ sgn } (r^*(k_1) - \zeta(k_1, \bar{w}^*, C)) & k_1 < k \le k_1 + d \\ r^*(k-1) & k > k_1 + d \end{cases} \quad (7.4.12)$$

and

$$w^*(k) = \begin{cases} 0 & k \le k_1 \\ w^*(k-1) + \delta_2 \text{ sgn } (f_{d+k_1-k}(r^*(k_1) - \zeta(k_1, \bar{w}^*, C))) & k_1 < k \le k_1 + d \\ w^*(k-1) & k > k_1 + d \end{cases}$$

$$(7.4.13)$$

where $k_1 \ge k_0$. It can be shown that $\bar{w}_E^* \in \mathbb{B}(\delta_1) \times \mathbb{B}(\delta_2)$. Substituting Eqs.(7.4.12) and (7.4.13) into the first equation of Eq.(7.4.5), we have

$$\begin{aligned} e(k_1 + d, \bar{w}^*, C) &= r^*(k_1) - \zeta(k_1, \bar{w}^*, C) + (d\delta_1 + \\ &+ \|F\|_1 \delta_2) \times \text{ sgn } (r^*(k_1) - \zeta(k_1, \bar{w}^*, C)), \end{aligned} \quad (7.4.14)$$

which also implies that

$$|e(k_1 + d, \bar{w}^*, C)| = |r^*(k_1) - \zeta(k_1, \bar{w}^*, C)| + d\delta_1 + \|F\|_1 \delta_2 \quad (7.4.15)$$

According to Eq.(7.4.2), it follows from Eq.(7.4.15) that

$$\phi_S(C) \ge \inf\{|r(k) - \zeta(k, \bar{w}, C)| : k \in \mathbb{N}\} + d\delta_1 + \|F\|_1 \delta_2. \quad (7.4.16)$$

By combinig Eqs.(7.4.11) and (7.4.16), the following inequality

$$\begin{aligned} \inf\{|r(k) - \zeta(k, \bar{w}, C)| : k \in \mathbb{N}\} &\le \\ &\le \phi_S(C) - (d\delta_1 + \|F\|_1 \delta_2) \le \\ &\le \sup\{|r(k) - \zeta(k, \bar{w}, C)| : k \in \mathbb{N}\}. \quad (7.4.17) \end{aligned}$$

is obtained. It can be seen that if $\zeta(k, \bar{w}, C) = r(k), \forall t \in \mathbb{N}$, $\phi_S(C)$ is minimized. Thus, with the initial setting $\zeta(k, \bar{w}, C) = r(k), \forall k \in \mathbb{N}$, the mean regulator $\bar{C}_O$ can be constructed. The optimal output performance is

$$\phi_S(\bar{C}_O) = d\delta_1 + \|F\|_1 \delta_2. \quad (7.4.18)$$

and Eq.(7.4.8) holds. □□

Theorem 7.2 shows that the mean controller $\bar{C}_O$ can limit the mean of the worst absolute error of the system with $\bar{w}_E \in \mathbb{B}(\delta_1) \times \mathbb{B}(\delta_2)$ to the bound $d\delta_1 + \|F\|_1 \delta_2$.

Example 7.2: Consider the plant of example 7.1 again, but now with disturbance ($d_1 = 0.05, \forall k$) and with the reference input $r(k)$ given by

$$\begin{aligned} r(k) &= r(k-1) + (0.02 + \alpha(k)) \text{ sgn } (\beta(k)) \\ &+ 0.02 \exp(-0.01t), \quad (7.4.19) \end{aligned}$$

where α and β are similar to ε and μ in Example 7.1, respectively. Clearly, $r_E \in \mathbb{B}(0.04)$ and $\omega_E \in \mathbb{B}(0.05)$. The mean controller $\bar{C}_O$ for the system is therefore given by

$$u(k) \;=\; r(k) + 0.2r(k-1) - 0.568y(k) - 0.506y(k-1) - 0.126y(k-2)$$

$$+u(k-1) - 0.54u(k-2) + 0.82u(k-3) - 0.28u(k-4) \qquad (7.4.20)$$

and

$$\phi_S(\bar{C}_O) = 0.225. \qquad (7.4.21)$$

The simulation results are shown in Figs. 7.3 and 7.4, where the behaviour of the system using the mean controller Eq.(7.4.20) is displayed. It can be seen that the mean controller $\bar{C}_O$ can guarantee that the mean of the absolute outputs of the system is limited to the prescribed bound, which is 0.225.

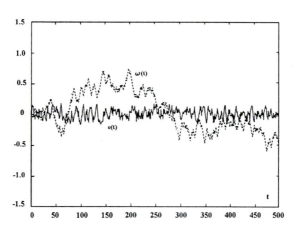

Figure 7.3: The disturbance of the system.

7.5 Stability of the Closed-Loop Systems

This section discusses the stability of the closed-loop systems with mean controllers. Let $r = 0$, then the following closed-loop system equations can be formulated from the plant Eq.(7.3.2) and the mean regulator Eq.(7.3.15)

$$y(k) = \frac{\Delta F B_n D_n}{\Delta F B_n B_d + E B_n q^{-d}} \omega(k). \qquad (7.5.1)$$

Using Eq.(7.3.3) it can be obtained that

$$y(k) = \frac{B_n D_n \Delta F}{B_n D_n} \omega(k). \qquad (7.5.2)$$

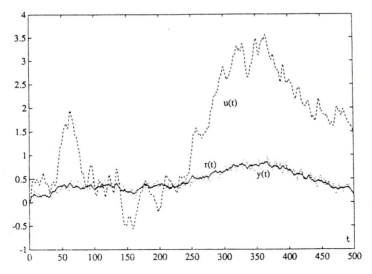

Figure 7.4: The output of the system.

Obviously, the characteristic equation of the closed-loop system is

$$B_n D_n = 0. \tag{7.5.3}$$

In the next stage, we consider the system with both the reference input r and the disturbance w. Using the plant Eq.(7.3.2) and the mean controller Eq.(7.4.6), the output y of the plant can be expressed as

$$y(k) = \frac{B_n D_n q^{-d} r(k) + \Delta F B_n D_n w(k)}{\Delta F B_n B_d + E B_n q^{-d}}. \tag{7.5.4}$$

As a result, the following closed-loop system equation can be formulated.

$$e(k) = \frac{(\Delta F B_n B_d + E B_n q^{-d} - B_n D_n q^{-d}) r(k) - \Delta F B_n D_n w(k)}{\Delta F B_n B_d + E B_n q^{-d}}. \tag{7.5.5}$$

With Eq.(7.3.3), we have

$$e(k) = \frac{B_n D_n}{B_n D_n} ((1 - q^{-d}) r(k) - \Delta F w(k)). \tag{7.5.6}$$

It can be seen that the characteristic equation of the closed-loop system is also

$$B_n D_n = 0. \tag{7.5.7}$$

If the polynomials B_n and D_n have all zeros outside the closed unit disc of the complex plane, then the closed-loop systems with the mean regulator $\bar{C}_R$ or the mean controller

$\bar{C}_O$ are stable and the term $B_n D_n$ in Eq.(7.5.2) and Eq.(7.5.6) can be cancelled without obscuring stability, so that for the system with the mean regulator $\bar{C}_R, y(k) = \Delta F \omega(k)$ and for the system with the mean controller $\bar{C}_O, e(k) = \rho(k) - \Delta F \omega(k)$. Of course, we can use the similar methods in Section 6.6 to design mean controllers if the polynomial B_n has some zeros within the closed unit disc.

7.6   The Self-Tuning Mean Regulator Algorithm

Consider a plant without reference input described by

$$B_d y(k) = q^{-d} B_n u(k) + D_n \omega(k),$$ (7.6.1)

where the polynomials $B_d \in \mathbb{R}[q, \bar{n}]$ with $b_{d0} = 1$, $B_n \in \mathbb{R}[q, \bar{m}]$ and $D_n \in \mathbb{R}[q, \bar{l}]$ with $D_{n0} = 1$, the time delay $d \in \mathbb{N}^+ \setminus 0$; y, u and ω are the output, the control and the disturbance of the system, respectively; $\omega_E \in \mathbb{B}(\delta)$.

For the plant described by Eq.(7.6.1), if the parameters of the plant are known the mean regulator of the form (7.3.15) can be directly obtained. In the case where the parameters are unknown we use an MLS identification algorithm for the stochastic systems to determine the parameters of the plant and then use the control law Eq.(7.3.15) with the true parameters substituted by their estimates. Following the MLS algorithm for the determinant case an MLS algorithm for the stochastic case is given in this section to estimate the parameters of the plant Eq.(7.6.1) with $\omega_E \in \mathbb{B}(\delta)$.

Again, we use y_k, u_k and ω_k to denote $y(k)$, $u(k)$ and $w(k)$, respectively, and

$$G = 1 - \Delta B_d,$$ (7.6.2)

where

$$G = g_1 q^{-1} + g_2 q^{-2} + ... + g_{\bar{n}+1} q^{-\bar{n}-1}.$$ (7.6.3)

Using these notations, the plant (7.6.1) can be expressed as

$$y_k = \theta^T \psi_{k-1} + \Delta \omega_k,$$ (7.6.4)

where

$$\theta = [g_1, g_2, ..., g_{\bar{n}+1}, b_{n0}, b_{n1}, ..., b_{n\bar{m}}, d_{n1}, d_{n2}, ..., d_{n\bar{l}}]^T,$$ (7.6.5)

$$\psi_{k-1} = [y_{k-1}, ..., y_{k-\bar{n}-1}, \Delta u_{k-d}, ..., \Delta u_{k-d-\bar{m}}, \Delta \omega_{k-1}, ..., \Delta \omega_{k-\bar{l}}]^T,$$ (7.6.6)

$$\Delta u_k = u_k - u_{k-1},$$ (7.6.7)

$$\Delta \omega_k = \omega_k - \omega_{k-1}.$$ (7.6.8)

It is clear that ψ_{k-1} can not be obtained because the sequence $\{\Delta\omega_{k-1}, \Delta\omega_{k-2}, ..., \Delta\omega_{k-\bar{l}}\}$ is unknown. Thus, we have to use ϵ_k to replace $\Delta\omega_k$, which leads to

$$\bar{\psi}_{k-1} = [y_{k-1}, ..., y_{k-\bar{n}-1}, \Delta u_{k-d}, ..., \Delta u_{k-d-\bar{m}}, \epsilon_{k-1}, ..., \epsilon_{k-\bar{l}}]^T \qquad (7.6.9)$$

Since the output posteriori estimate is $\bar{y}_k = \hat{\theta}_k^T \bar{\psi}_{k-1}$, ϵ_k is defined as the posteriori output error, i.e.

$$\epsilon_k = y_k - \bar{y}_k \qquad (7.6.10)$$

The purpose of the estimation is then to find a θ belonging to the set defined by

$$\Gamma(\theta) = \{\theta : E\{|y_k - \theta^T \bar{\psi}_{k-1}|\} \le \delta, \forall k \in \mathbb{N}\}. \qquad (7.6.11)$$

Similar to the MLS algorithm for the determinant systems in Section 6.7, the following MLS algorithm for stochastic systems can be formulated.

Theorem 7.3 *Consider the plant (7.6.1) and the identification algorithm:*

$$\hat{\theta}_k = \hat{\theta}_{k-1} + \frac{\alpha_k P_{k-1} \bar{\psi}_{k-1}}{1 + \bar{\psi}_{k-1}^T P_{k-1} \bar{\psi}_{k-1}} (|e_k| - \delta) \text{ sgn } (e_k), \qquad (7.6.12)$$

$$P_k = P_{k-1} - \frac{\alpha_k P_{k-1} \bar{\psi}_{k-1} \bar{\psi}_{k-1}^T P_{k-1}}{|e_k| + (2|e_k| - \delta)\bar{\psi}_{k-1}^T P_{k-1}\bar{\psi}_{k-1}} (|e_k| - \delta), \qquad (7.6.13)$$

$$e_k = y_k - \hat{\theta}_{k-1}^T \bar{\psi}_{k-1}, \qquad (7.6.14)$$

$$\alpha_k = \begin{cases} 1, & |e_k| > \delta \\ 0, & |e_k| \le \delta \end{cases} \qquad (7.6.15)$$

Then

$$i) \qquad \lim_{k \to \infty} \frac{\alpha_k(|e_k| - \delta)^2}{1 + \bar{\psi}_{k-1}^T P_{k-1}\bar{\psi}_{k-1}} = 0, \quad a.s. \qquad (7.6.16)$$

$$ii) \qquad \lim_{k \to \infty} |\hat{\theta}_k - \hat{\theta}_{k-1}| = 0, \quad a.s. \qquad (7.6.17)$$

$$iii) \qquad \left\|\tilde{\theta}_k\right\|^2 \le \frac{\lambda_{max}(P_0^{-1})}{\lambda_{min}(P_0^{-1})} \left\|\tilde{\theta}_0\right\|^2, \quad a.s. \qquad (7.6.18)$$

where

$$\tilde{\theta}_k = \theta - \hat{\theta}_k \qquad (7.6.19)$$

and $\lambda_{max}(.)$ and $\lambda_{min}(.)$ denote the maximum and the minimum eigenvalues of the matrix (.).

The proof of the theorem is similar to Theorem 6.6. In order to control a system with constant but unknown parameters and with $\omega_E \in \mathbb{B}(\delta)$ a self-tuning mean regulator algorithm is given below which is based on the mean regulator developed in Section 7.3 and the MLS algorithm in theorem 7.3. In parallel to assumption 6.1, we have

Assumption 7.1

    i) The time delay d is known.

    ii) An upper bound for the orders of the polynomials B_d, B_n and D_n in the plant (7.6.1) is known.

    iii) The polynomials B_n and D_n have all zeros outside the closed unit disc, i.e., stable polynomials.

The self-tuning mean regulator algorithm is as follows:

  i) **Step 1: Parameter Estimation**

At the sampling interval k determine the parameter θ satisfying

$$E\{|y_k - \theta^T \bar{\psi}_{k-1}|\} \le \delta, \forall k \in \mathbb{N} \tag{7.6.20}$$

and then compute the polynomials F and E by solving the following equation

$$\hat{F}_k(1 - \hat{G}_k) + \hat{E}_k q^{-d} = \hat{D}_{nk}, \tag{7.6.21}$$

where $\hat{F}_k$, $\hat{G}_k$, $\hat{E}_k$ and $\hat{D}_{nk}$ are the estimated values of the polynomials F, G, E and D_n at time k, respectively, and $\hat{f}_0(k) = 1$.

  ii) **Step 2: Control**

The objective is to design a control law such that the output of the system satisfies

$$\limsup_{k \to \infty} |y_k| \le \left\| \hat{F} \right\|_1 \delta, \quad a.s. \tag{7.6.22}$$

where

$$\hat{F} = \lim_{k \to \infty} \hat{F}_k. \tag{7.6.23}$$

At each sampling interval determine the control variable from the mean regulator $\bar{C}_R^k$:

$$\hat{E}_k y_k + \hat{F}_k \hat{B}_{nk} \Delta u(k) = 0 \tag{7.6.24}$$

where the polynomials $\hat{E}_k$, $\hat{F}_k$ and $\hat{B}_{nk}$ are obtained in Step 1.

Since the estimation algorithm in this section is computed recursively, the algorithm requires only moderate computations. For the analysis of the convergence of the self-tuning mean regulator algorithm, the following global convergence result of the self-tuning mean regulator algorithm is established.

Theorem 7.4 *Subject to Assumption 7.1, the self-tuning mean regulator algorithm (7.6.12)-(7.6.15) and (7.6.24), when applied to the plant (7.6.1), yields*

$$
\begin{array}{lll}
i) & E\{y(k)\} \text{ and } E\{u(k)\} \text{ are bounded sequences.} & (7.6.25) \\
ii) & \lim_{k\to\infty} \sup |e_k| \leq \delta, \quad a.s. & (7.6.26) \\
iii) & \lim_{k\to\infty} \sup |y_k| \leq \left\|\hat{F}\right\|_1 \delta, \quad a.s. & (7.6.27)
\end{array}
$$

Using the similar procedure as we have used in proving theorem 6.7, theorem 7.4 can be easily proved. It can seen from this theorem that the closed-loop stability is achieved and the output of the system is asymptotically limited to the bounds $\pm\left\|\hat{F}\right\|_1 \delta$ a.s. The performance of the self-tuning mean regulator scheme is demonstrated by the following example.

Example 7.3 Let the plant be

$$(1 + 1.9q^{-1} + 0.92q^{-2})y(k) = q^{-1}(1 + 0.85q^{-1})u(k) + \omega(k), \tag{7.6.28}$$

where the disturbance is chosen as

$$\omega(k) = \omega(k-1) + 0.1\nu(k) \tag{7.6.29}$$

and $\nu(k)$ is a random variable with $E\{\nu(k)\} = 0$ and $E\{\nu^2(k)\} = 1$. From Eq.(7.2.3), it can be seen that the disturbance $\omega \in \mathbb{B}(\delta)$ with $\delta = 0.1$.

Using theorem 7.1, the mean regulator C_R of the form

$$u(k) = 0.9y(k) - 0.98y(k-1) - 0.92y(k-2) + 0.15u(k-1) + 0.85u(k-2) \tag{7.6.30}$$

can be obtained and the optimal output performance

$$\phi_S(C_R) = 0.1. \tag{7.6.31}$$

From Eq.(7.6.21), it follows that

$$\hat{F}_k = 1, \quad \hat{E}_k = (1 - \hat{G}_k)q. \tag{7.6.32}$$

The regulator with the same structure as Eq.(7.6.30) is obtained using the self-tuning algorithm in this section. Using Eqs.(7.3.16) and (7.6.27), it can be shown that

$$\lim_{k\to\infty} \sup \phi_{SR}(C_R^k) \leq 0.1, \quad a.s. \tag{7.6.33}$$

The output $y(k)$ and the parameter error norm $\left\|\theta - \hat{\theta}_k\right\|^2$ are shown in Figures 7.5 and 7.6, respectively.

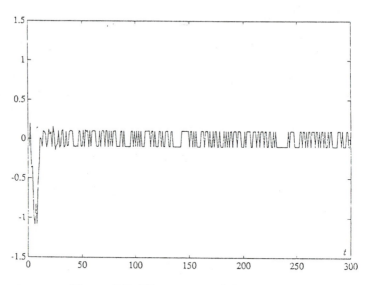

Figure 7.5: The output of the system.

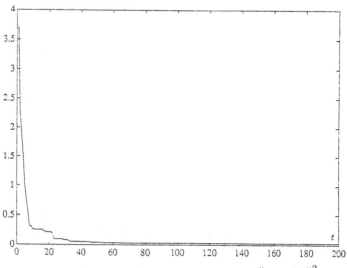

Figure 7.6: The parameter error norm $\left\| \theta - \hat{\theta}_k \right\|^2$.

7.7   Conclusions

In parallel to chapter 6, the design procedure for the control of stochastic systems, which is subjected to random inputs, is discussed. In contrast to classical adaptive control where external inputs (or disturbances) are assumed to be white noises with finite variances, the disturbances in this chapter are modelled by a random input space characterized by a bound on the mean of the absolute increment of the external inputs. The performance function for the systems is then defined as the supremum of the mean of the absolute value of the generalized output for all external inputs in the random input space and for all time. A controller, which minimizes this performance function and is referred to as mean controller, is developed. Again, for the systems without reference input, the mean controller becomes the mean regulator. Some fundamental relationships between random input space, output performance and controller are established.

The self-tuning control for unknown systems is also considered and a self-tuning mean regulator algorithm is obtained via the use of the MLS estimate for the stochastic case and the structure of the mean regulator. The main result on the characterization of the closed-loop system is obtained when the self-tuning mean regulator algorithm is applied to the system. It has been shown that the properties of the self-tuning mean regulator will converge to those of mean regulators.

Chapter 8

MRAPC for Time Delay Systems

8.1 Introduction

It is well known that time delay widely exists in many industrial processes. Typical examples are heat flow, material transportation, chemical reactors and distillation columns. The difficulties caused by the presence of the time delay in control systems have been recognized for many years. For systems in which the time delay is significant and where high performance is required, conventional controllers, such as PI and PID, are difficult to apply. This is mainly due to the additional phase lag introduced by the time delay which tends to destabilize closed-loop systems. To compensate this effect, the gain of controller is usually reduced. Although the reduction of the gain may solve the problem of stability, it results in sluggish closed-loop response. In 1957, Smith first proposed a predictor control (SPC) scheme [141, 123], which effectively removed the time delay from the closed-loop characteristic equation. The basic idea of the SPC is to use a delay-free model of the rational dynamics of the process to predict the effect of current control action on the actual delayed process output. Additional feedback from the actual process output is then used to compensate modelling errors and process disturbances. As a result, SPC allows the use of well developed classical design techniques for rational transfer functions presented systems. Further studies on SPC [20, 59, 68, 123] have shown its improved performance over conventional controllers and its high sensitivity to modelling errors. It has been found that the system performs well if the process model is accurate, and that the performance degrades when the inaccuracy in the process parameters and the time delay gets serious. However, the requirement on an accurate modelling is very strict for many industrial processes, where the parameters cannot be easily estimated or even change their values after being estimated. To improve the performance of the SPC, an additional regulator is added to the SPC [30, 42, 43, 60, 61, 79, 160], which is called the Modified Smith Predictor Control (MSPC). Although the MSPC is better than the SPC in reducing the influence of the modelling error, it is still sensitive to the parameter variations of the process, and the most sensitive one is the time delay of the system.

To realize adaptive control for systems with unknown constant or variable time delay, two approximations are used to replace the time delay transfer function e^{-sL} in [29, 53] as follows :

$$D_r(s) = \frac{1}{\left(1 + \dfrac{sL}{r}\right)^r}, \qquad D_p(s) = \left(\frac{1 - \dfrac{sL}{p}}{1 + \dfrac{sL}{p}}\right)^p, \qquad (8.1.1)$$

where L is the time delay, r and p are the orders of the approximations.

Based on the structure of Smith predictor control strategy, the parametric optimization theory for adaptive control and the Lyapunov stability theory, this chapter describes a model reference adaptive predictor and control (MRAPC) scheme for time-varying systems with variable time delay.

8.2 The Smith Predictive Control

The Smith predictor control (SPC) scheme, which was first proposed by Smith [141, 142], uses a delay-free model of the rational dynamics of the process to predict the effects of current control action on the actual delayed process output. A feedback from the actual process output is then used to compensate for modelling errors and process disturbances. Hence, the SPC effectively removes the time-delay from the closed-loop characteristic equation for systems with time delay. It also allows the use of classical design techniques developed for rational transfer functions.

Figure 8.1 shows the structure of Smith predictor control scheme, where block $G_m(s)$ $(e^{-sL_m} - 1)$ is the Smith predictor, whilst $G_p(s)e^{-sL_p}$ and $G_c(s)$ represent the plant and the conventional controller, respectively. The closed-loop transfer function for such a system is given by

$$G(s) = \frac{G_c(s)G_p(s)e^{-sL_p}}{1 + G_c(s)G_m(s) + G_c(s)(G_p(s)e^{-sL_p} - G_m(s)e^{-sL_m})} \qquad (8.2.1)$$

It can be seen that the stability of the Smith predictor is affected by the accuracy of the model which represents the plant. Based on the assumption that the model used perfectly matches the plant dynamics, i.e. $G_p(s) = G_m(s)$ and $L_p = L_m$, the closed-loop transfer function can be reduced to give

$$G(s) = \frac{G_c(s)G_p(s)e^{-sL_p}}{1 + G_c(s)G_m(s)} \qquad (8.2.2)$$

Clearly, the characteristic polynomial of the closed-loop system, $1 + G_c(s)G_m(s)$, does not contain e^{-sL_p} or e^{-sL_m}, and the controller $G_c(s)$ can be designed as if the controlled process is $G_p(s)$. The desired response to the reference input can then be obtained.

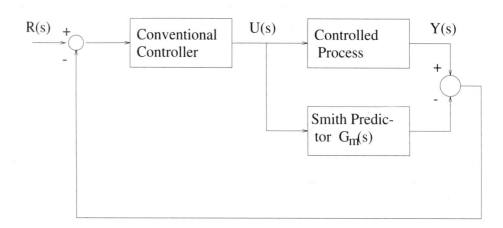

Figure 8.1: The Smith predictor control scheme.

In practice, however, the perfect matching never exists. The parameters and the time delay of the model always differ from those of the pant, i.e. $G_m(s) \neq G_p(s)$ and $L_m \neq L_p$, Therefore, there exist terms e^{-sL_p} and e^{-sL_m} in the characteristic polynomial of the closed-loop system. As a result, the number of poles of the closed-loop system will increase due to the presence of e^{-sL_p} and e^{-sL_m}. These added poles may result in an unstable closed-loop system. We can decrease the gain of the controller $G_c(s)$ to reduce the influence of e^{-sL_p} and e^{-sL_m}, but the decrease of the gain may lead to the zeroes of $1 + G_c(s)G_m(s)$ too close to the imaginary axis $j\omega$. For example, if we choose

$$G_c(s) = K_c, \qquad G_m(s) = \frac{K_m}{1 + sT_m}, \qquad (8.2.3)$$

then, the zero of $1 + G_c(s)G_m(s)$ is $s = -(K_m K_c + 1)/T_m$. Therefore, an originally stable system may become either unstable, or oscillate, or even converge slowly when the parameters or the time delay of the model does not match those of the controlled plant.

8.3   The Modified Smith Predictor Control

To improve the performance of SPC, a Modified Smith Predictor Control (MSPC) scheme is proposed and its structure is shown in Figure 8.2, where $G_m(s)e^{-sL_m}$ represents the dynamic model of the plant, $G_p(s)e^{-sL_p}$ and $G_c(s)$ are the transfer functions of the plant and the conventional controller, respectively. M(s) is a dynamical compensator designed to eliminate the effects caused by external disturbances. The transfer function between the reference input $R(s)$ and the output $Y(s)$ is therefore given by

$$G(s) = \frac{G_c(s)G_p(s)e^{-sL_p}}{1 + G_c(s)G_m(s) + G_c(s)M(s)(G_p(s)e^{-sL_p} - G_m(s)e^{-sL_m})}. \qquad (8.3.1)$$

In the case where the plant and the model perfectly match, i.e., $G_p(s) = G_m(s)$ and $L_p = L_m$, the characteristic polynomial of the closed-loop system, again, does not contain e^{-sL_p} or e^{-sL_m}.

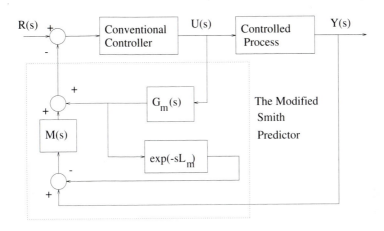

Figure 8.2: The modified Smith predictor control scheme.

In practice, the parameters and the time delay of the plant will always differ from those of the model, as a result, terms e^{-sL_p} and e^{-sL_m} will appear in the characteristic polynomial of the closed-loop system, which is given by $1 + G_c(s)G_m(s) + G_c(s)M(s)(G_p(s)e^{-sL_p} - G_m(s)e^{-sL_m})$. It can be seen that $M(s)$ can be designed to reduce the influence of $G_p(s)e^{-L_p s} - G_m(s)e^{-L_m s}$ to some certain extent for small variations in the process parameters, but the reduction of such an influence is rather limited.

8.4 MRAPC Using Parametric Optimization Theory

Using the Smith predictor control and the parametric optimization theory, this section presents a Model Reference Adaptive Predictor Control (MRAPC) for time-varying systems with variable time delay [88, 89, 93]. The MRAPC consists of an adaptive plant predictor, an adaptive process model, an adaptation mechanism and a conventional controller (see Fig. 8.3). The predictor is used to estimate the effect of the current control action on the actual plant output, whilst the conventional controller ensures that the system will have a desired closed loop performance.

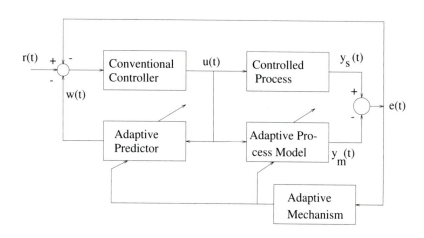

Figure 8.3: The model reference adaptive predictor control scheme.

Define the generalized error as

$$e(t) = y_s(t) - y_m(t) \qquad (8.4.1)$$

where $y_s(t)$ and $y_m(t)$ are the outputs of the plant and model, respectively, and denote

$$J(t) = \frac{1}{2} \int_{t_0}^{t} e^2(t)dt. \qquad (8.4.2)$$

as a performance index function, then it can be seen that the controlled plant can be expressed as

$$\sum_{i=0}^{N} \bar{a}_i(t)p^i y_s(t) = \sum_{j=0}^{M} \bar{b}_j(t)p^j u(t - \bar{L}_p(t)) \qquad (8.4.3)$$

where $u(t)$ is the control input, $\bar{a}_i(t)$ $(i = 0, 1, ..., N)$ and $\bar{b}_j(t)$ $(j = 0, 1, ..., M)$ are time-varying parameters, $\bar{L}_p(t)$ is a variable time delay and $p = \frac{d}{dt}$ is the differential operator.
The process model is given by

$$\sum_{i=0}^{N} a_i(e, t)p^i y_m(t) = \sum_{j=0}^{M} b_j(e, t)p^j u(t - L_m(e, t)) \qquad (8.4.4)$$

where $a_i(e, t)$ $(i = 0, 1, ..., N)$ and $b_j(e, t)$ $(j = 0, 1, ..., M)$ are adjustable parameters, and $L_m(e, t)$ is an adjustable time delay.

By removing the time delay $L_m(e,t)$ from Eq. (8.4.4), the adaptive predictor of the form

$$\sum_{i=0}^{N} a_i(e,t)p^i w(t) = \sum_{j=0}^{M} b_j(e,t)p^j u(t), \qquad (8.4.5)$$

can be readily obtained, where $w(t)$ is the output of the predictor. In order to obtain an approximative optimization solution so that the adaptive control law is simple and easily realizable in practice, we assume that

i) the controlled process is slowly time-varying;

ii) the variation ranges of the controlled process parameters are not very large;

iii) the updating rate of the adaptation mechanism is slow.

Without loss of generality we further assume that $\bar{a}_0(t) = a_0(e,t) = 1$, $\bar{a}_N(t) \neq 0$ and $a_N(e,t) \neq 0$. Under these assumptions, the following adaptation mechanism can be obtained via the use of the parametric optimization theory.

$$a_i(e,t) = -\lambda_{a_i} \frac{\partial J(t)}{\partial a_i} + a_{i0} \qquad (8.4.6)$$

$$b_j(e,t) = -\lambda_{b_j} \frac{\partial J(t)}{\partial b_j} + b_{j0} \qquad (8.4.7)$$

$$L_m(e,t) = -\lambda_{Lm} \frac{\partial J(t)}{\partial L_m} + L_{m0} \qquad (8.4.8)$$

where λ_{a_i}, λ_{b_j} and λ_{Lm} are arbitrary positive constants, a_{i0}, b_{j0} and L_{m0} are initial values of $a_i(e,t)$, $b_j(e,t)$ and $L_m(e,t)$, respectively, and $i = 1,2,...,N$ and $j = 0,1,...,M$. Using Eqs. (8.4.1) and (8.4.2), it can be obtained that

$$\frac{\partial a_i(e,t)}{\partial t} = -\lambda_{a_i} e(t) \frac{\partial e(t)}{\partial a_i} = \lambda_{a_i} e(t) \frac{\partial y_m(t)}{\partial a_i}, \qquad (8.4.9)$$

$$\frac{\partial b_j(e,t)}{\partial t} = -\lambda_{b_j} e(t) \frac{\partial e(t)}{\partial b_j} = \lambda_{b_j} e(t) \frac{\partial y_m(t)}{\partial b_j}, \qquad (8.4.10)$$

$$\frac{\partial L_m(e,t)}{\partial t} = -\lambda_{Lm} e(t) \frac{\partial e(t)}{\partial L_m} = \lambda_{Lm} e(t) \frac{\partial y_m(t)}{\partial L_m} \qquad (8.4.11)$$

where $i = 1,2,...,N$ and $j = 0,1,...,M$. To implement the adaptation mechanism, we must generate the sensitivity functions

$$\frac{\partial y_m(t)}{\partial a_i}, \quad \frac{\partial y_m(t)}{\partial b_j}, \quad \frac{\partial y_m(t)}{\partial L_m} \qquad (8.4.12)$$

for $i = 1, 2, ..., N$ and $j = 0, 1, ..., M$. With the assumptions i) and iii), and by taking the partial differential in both sides of Eq.(8.4.4) with respect to a_i $(i = 1, 2, ..., N)$, b_j $(j = 0, 1, ..., M)$, and L_m, respectively, and by exchanging the order of differentials, we have

$$\frac{\partial y_m(t)}{\partial a_i} \approx -p^i y_m(t) - \sum_{k=1}^{N} a_k(e,t) p^k \frac{\partial y_m(t)}{\partial a_i}, \tag{8.4.13}$$

$$\frac{\partial y_m(t)}{\partial b_j} \approx p^j u(t - L_m(e,t)) - \sum_{k=1}^{N} a_k(e,t) p^k \frac{\partial y_m(t)}{\partial b_j}, \tag{8.4.14}$$

$$\frac{\partial y_m(t)}{\partial L_m} \approx \sum_{k=1}^{M} b_k(e,t) p^k \frac{\partial u(t - L_m(e,t))}{\partial L_m} - \sum_{k=1}^{N} a_k(e,t) p^k \frac{\partial y_m(t)}{\partial L_m} \tag{8.4.15}$$

for $i = 1, 2, ..., N$ and $j = 0, 1, ..., M$. Using the assumptions i) and iii), it can be further obtained that

$$\frac{\partial y_m(t)}{\partial a_i} \approx \frac{\partial}{\partial t} \frac{\partial y_m(t)}{\partial a_{i-1}} \approx \frac{\partial}{\partial t^{i-1}} \frac{\partial y_m(t)}{\partial a_1} \tag{8.4.16}$$

for $i = 2, 3, ..., N$, and

$$\frac{\partial y_m(t)}{\partial b_j} \approx \frac{\partial}{\partial t} \frac{\partial y_m(t)}{\partial b_{j-1}} \approx \frac{\partial}{\partial t^j} \frac{\partial y_m(t)}{\partial b_0} \tag{8.4.17}$$

for $j = 1, 2, ..., M$. Based on the assumption ii), we can replaces $a_i(e,t)$ and $L_m(e,t)$ with a_{k0} and L_{m0} $(k = 1, 2, ..., N)$ in Eqs.(8.4.13) and (8.4.14). Therefore we have

$$\frac{\partial y_m(t)}{\partial a_1} \approx -p y_m(t) - \sum_{k=1}^{N} a_{k0} p^k \frac{\partial y_m(t)}{\partial a_1} \tag{8.4.18}$$

$$\frac{\partial y_m(t)}{\partial b_0} \approx u(t - L_{m0}(t)) - \sum_{k=1}^{N} a_{k0} p^k \frac{\partial y_m(t)}{\partial b_0} \tag{8.4.19}$$

Substituting Eq.(8.4.16) into Eq.(8.4.18), the following approximation can be obtained

$$\frac{\partial y_m(t)}{\partial a_1} \approx -p y_m(t) - \sum_{k=1}^{N-1} a_{k0} \frac{\partial y_m(t)}{\partial a_{k+1}} - a_{N0} p \frac{\partial y_m(t)}{\partial a_N} \tag{8.4.20}$$

therefore

$$\frac{\partial}{\partial t} \frac{\partial y_m(t)}{\partial a_N} \approx -\frac{1}{a_{N0}} \left[p y_m(t) + \frac{\partial y_m(t)}{\partial a_1} + \sum_{k=1}^{N-1} a_{k0} \frac{\partial y_m(t)}{\partial a_{k+1}} \right] \tag{8.4.21}$$

In the light of Eqs.(8.4.16)-(8.4.21), we have

$$\frac{\partial S(t)}{\partial t} \approx AS(t) + B\frac{\partial y_m(t)}{\partial t} \tag{8.4.22}$$

where

$$S(t) = \left[\frac{\partial y_m(t)}{\partial a_1}, \frac{\partial y_m(t)}{\partial a_2}, ..., \frac{\partial y_m(t)}{\partial a_N}\right]^T \tag{8.4.23}$$

$$A = \begin{bmatrix} 0 & 1 & 0 & \cdots & 0 \\ 0 & 0 & 1 & & 0 \\ \vdots & \vdots & & \ddots & \\ 0 & 0 & 0 & & 1 \\ -\frac{1}{a_{N0}} & -\frac{a_{10}}{a_{N0}} & -\frac{a_{20}}{a_{N0}} & \cdots & -\frac{a_{(N-1)0}}{a_{N0}} \end{bmatrix} \tag{8.4.24}$$

$$B = \begin{bmatrix} 0, & 0, & \cdots & 0, & -\frac{1}{a_{N0}} \end{bmatrix}^T \tag{8.4.25}$$

The solution of Eq.(8.4.22) is

$$S(t) \approx e^{A(t-t_0)}(S(t_0) - By_m(t_0)) + By_m(t) + \int_{t_0}^t e^{A(t-\tau)}ABy_m(\tau)d\tau \tag{8.4.26}$$

where $S(t_0)$ and $y_m(t_0)$ are the initial values of $S(t)$ and $y_m(t)$. In general, we can take $S(t_0) = 0$ and $y_m(t_0) = 0$, as a result

$$S(t) \approx By_m(t) + \int_{t_0}^t e^{A(t-\tau)}ABy_m(\tau)d\tau \tag{8.4.27}$$

Since $M < N - 1$, we can add $N - M - 1$ auxiliary parameters $b_{M+1}, b_{M+2}, ..., b_N$ such that the equalities

$$\frac{\partial y_m(t)}{\partial b_{M+i}} = \frac{\partial}{\partial t}\frac{\partial y_m(t)}{\partial b_{M+i-1}} \tag{8.4.28}$$

hold for $i = 1, 2, ..., N-M-1$. By substituting Eqs.(8.4.17) and (8.4.28) into Eq.(8.4.19), it can be seen that

$$\frac{\partial y_m(t)}{\partial b_0} \approx u(t - L_{m0}) - \sum_{k=1}^{N-1} a_{k0}\frac{\partial y_m(t)}{\partial b_k} - a_{N0}p\frac{\partial y_m(t)}{\partial b_{N-1}} \tag{8.4.29}$$

as a result, we have

$$\frac{\partial}{\partial t}\frac{\partial y_m(t)}{\partial b_{N-1}} \approx \frac{1}{a_{N0}}u(t - L_{m0}) - \sum_{k=1}^{N-1}\frac{a_{k0}}{a_{N0}}\frac{\partial y_m(t)}{\partial b_k} - \frac{1}{a_{N0}}\frac{\partial y_m(t)}{\partial b_0} \tag{8.4.30}$$

In terms of Eqs.(8.4.17), (8.4.28) and (8.4.30), it can be further obtained that

$$\frac{\partial M(t)}{\partial t} \approx CM(t) + Du(t - L_{m0}) \tag{8.4.31}$$

where

$$M(t) = \left[\frac{\partial y_m(t)}{\partial b_0}, \frac{\partial y_m(t)}{\partial b_1}, ..., \frac{\partial y_m(t)}{\partial b_{N-1}}\right]^T \tag{8.4.32}$$

$$C = \begin{bmatrix} 0 & 1 & 0 & \cdots & 0 \\ 0 & 0 & 1 & & 0 \\ \vdots & \vdots & & \ddots & \\ 0 & 0 & 0 & & 1 \\ -\frac{1}{a_{N0}} & -\frac{a_{10}}{a_{N0}} & -\frac{a_{20}}{a_{N0}} & \cdots & -\frac{a_{(N-1)0}}{a_{N0}} \end{bmatrix} \tag{8.4.33}$$

$$D = [0, \quad 0, \quad \cdots \quad 0, \quad \frac{1}{a_{N0}}]^T \tag{8.4.34}$$

Similar to Eq. (8.4.26), the solution for Eq.(8.4.31) can be expressed as

$$M(t) \approx e^{C(t-t_0)} M(t_0) + \int_{t_0}^{t} e^{C(t-\tau)} Du(\tau - L_{m0}) d\tau \tag{8.4.35}$$

where $M(t_0)$ is the initial values of $M(t)$, Since $M(t_0) = 0$, the following approximation can also be obtained

$$M(t) \approx \int_{t_0}^{t} e^{C(t-\tau)} Du(\tau - L_{m0}) d\tau \tag{8.4.36}$$

Using assumption iii), it can be shown that

$$p \sum_{k=0}^{N} a_k(e,t) p^k \approx \sum_{k=0}^{N} a_k(e,t) p^k p \tag{8.4.37}$$

$$p \sum_{k=0}^{M} b_k(e,t) p^k \approx \sum_{k=0}^{M} b_k(e,t) p^k p \tag{8.4.38}$$

$$\frac{\partial u(t - L_m(e,t))}{\partial L_m} \approx -\frac{\partial u(t - L_m(e,t))}{\partial t} \tag{8.4.39}$$

By taking partial derivatives with respect to t in both sides of the Eq.(8.4.4), we have

$$p \sum_{k=0}^{N} a_k(e,t) p^k y_m(t) = p \sum_{k=0}^{M} b_k(e,t) p^k u(t - L_m(e,t)) \tag{8.4.40}$$

Substituting Eqs.(8.4.37) and (8.4.38) into the above equation, the following approximation is obtained

$$\sum_{k=0}^{N} a_k(e,t)p^k p y_m(t) \approx \sum_{k=0}^{M} b_k(e,t)p^k p u(t - L_m(e,t)) \qquad (8.4.41)$$

Using Eq.(8.4.39), it can be seen from Eq.(8.4.15) that

$$\sum_{k=0}^{N} a_k(e,t)p^k \frac{\partial y_m(t)}{\partial L_m} \approx -\sum_{k=0}^{M} b_k(e,t)p^k p u(t - L_m(e,t)) \qquad (8.4.42)$$

Comparing Eq.(8.4.41) with Eq.(8.4.42), it can be concluded that

$$\frac{\partial y_m(t)}{\partial L_m} \approx -p y_m(t) \qquad (8.4.43)$$

At this stage, all sensitive functions are obtained from Eqs.(8.4.27), (8.4.36) and (8.4.43).
Therefore, in the light of Eqs.(8.4.9)-(8.4.11), we have

$$\begin{bmatrix} a_1(e,t) \\ a_2(e,t) \\ \vdots \\ a_N(e,t) \end{bmatrix} \approx \begin{bmatrix} \lambda_{a_0} & & & \\ & \lambda_{a_1} & & \\ & & \ddots & \\ & & & \lambda_{a_N} \end{bmatrix} \int_{t_0}^{t} S(t)e(t)dt + \begin{bmatrix} a_{10} \\ a_{20} \\ \vdots \\ a_{N0} \end{bmatrix} \qquad (8.4.44)$$

$$\begin{bmatrix} b_0(e,t) \\ b_1(e,t) \\ \vdots \\ b_M(e,t) \end{bmatrix} \approx \begin{bmatrix} \lambda_{b_1} & & & \\ & \lambda_{b_2} & & \\ & & \ddots & \\ & & & \lambda_{b_M} \end{bmatrix} \int_{t_0}^{t} [I_m \vdots O] M(t)e(t)dt + \begin{bmatrix} b_{00} \\ b_{10} \\ \vdots \\ b_{M0} \end{bmatrix} \qquad (8.4.45)$$

$$L_m(e,t) \approx -\lambda_{Lm} \int_{t_0}^{t} \dot{y}_m(t)e(t)dt + L_{m0} \qquad (8.4.46)$$

where $I_m = \text{diag}(1,1,...,1) \in \mathbb{R}^{(M+1)\times(M+1)}$ and $O = 0 \in \mathbb{R}^{(M+1)\times(N-M-1)}$. Eqs.(8.4.44)-
(8.4.46) are the adaptive laws of the parameters $a_i(e,t)$ $(i = 1,2,...,N)$, $b_j(e,t)$ $(j = 0,1,..,M)$ and the time delay $L_m(e,t)$.

Example 8.1 Consider the following first order dynamic system with time delay

$$(\bar{a}_1(t)p + 1)y_s(t) = \bar{b}_0(t)u(t - \bar{L}_p(t)) \qquad (8.4.47)$$

with the nominal value of $\bar{a}_1(t)$, $\bar{b}_0(t)$ and $\bar{L}_p(t)$ being set to 40, 5 and 4.5, respectively.
The process model in this case can be selected as

$$(a_1(e,t)p + 1)y_m(t) = b_0(e,t)u(t - L_m(e,t)) \qquad (8.4.48)$$

and the corresponding predictor is therefore given by

$$(a_1(e,t)p + 1)w(t) = b_0(e,t)u(t) \tag{8.4.49}$$

The conventional controller for this example is PID and is given by

$$G_c(s) = 4.7(1 + \frac{1}{16s} + 0.16s) \tag{8.4.50}$$

The adaptive laws of the parameters in the process model and the predictor can be obtained from Eqs.(8.4.44)-(8.4.46), where a_{10}, b_{00}, and L_{m0} are chosen to be 40, 5 and 4.5. The sampling period is set at 0.5 seconds and the demand signal, $y_r(t)$, is a square wave of period 60 seconds. The simulation results are shown in Figs. 8.4 and 8.5, where Figures 8.4(a) and 8.5(a) display the dynamic performances of the system using the adaptive control laws whilst Figures 8.4(b) and 8.5(b) give the dynamic performances of the system using standard Smith Predictor Control.

To simulate parameter variations, in Fig. 8.4 the parameters $\bar{a}_1(t), \bar{b}_0(t)$ and $\bar{L}_p(t)$ are changed as follows:

$$
\begin{align}
t \in [0, 30] \quad &\bar{a}_1(t) = 40, \quad \bar{b}_0(t) = 5, \quad \bar{L}_p(t) = 4.5, \tag{8.4.51}\\
t \in [30, 120) \quad &\bar{a}_1(t) = 43, \quad \bar{b}_0(t) = 5.5, \quad \bar{L}_p(t) = 5, \tag{8.4.52}\\
t \in [120, 210) \quad &\bar{a}_1(t) = 46, \quad \bar{b}_0(t) = 6, \quad \bar{L}_p(t) = 5.5, \tag{8.4.53}\\
t \in [210, 300] \quad &\bar{a}_1(t) = 50, \quad \bar{b}_0(t) = 6.5, \quad \bar{L}_p(t) = 6. \tag{8.4.54}
\end{align}
$$

whilst in Fig. 8.5, the parameters $\bar{a}_1(t), \bar{b}_0(t)$ and $\bar{L}_p(t)$ are subjected to

$$
\begin{align}
t \in [0, 30] \quad &\bar{a}_1(t) = 40, \quad \bar{b}_0(t) = 5, \quad \bar{L}_p(t) = 4.5, \tag{8.4.55}\\
t \in [30, 120) \quad &\bar{a}_1(t) = 37, \quad \bar{b}_0(t) = 4.5, \quad \bar{L}_p(t) = 5, \tag{8.4.56}\\
t \in [120, 210) \quad &\bar{a}_1(t) = 34, \quad \bar{b}_0(t) = 4, \quad \bar{L}_p(t) = 5.5, \tag{8.4.57}\\
t \in [210, 300] \quad &\bar{a}_1(t) = 30, \quad \bar{b}_0(t) = 3.5, \quad \bar{L}_p(t) = 6. \tag{8.4.58}
\end{align}
$$

It can be seen that the standard Smith predictor cannot cope with large parameter variations whilst the MRAPC can still lead to a very good closed-loop performance.

8.5   MRAPC Using Lyapunov Stability Theory

Although the method discussed in sections 8.2-8.4 can be used to generate a desired performance, it cannot guarantee the stability of the closed loop system. In this section, an alternative approach [96] is presented, where a Lyapunov function is used to construct a stable control strategy for time-varying delayed systems.

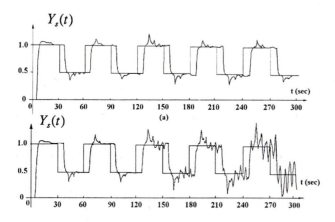

Figure 8.4: MRAPC simulation results (1).

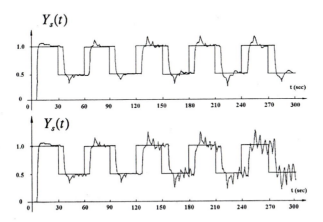

Figure 8.5: MRAPC simulation results (2).

8.5.1   The model reference based identification

Unlike section 8.4, we assume that the plant is expressed in a state space form as

$$\dot{x}_s(t) \;=\; A_s x_s(t) + B_s u(t - L_s(t)), \tag{8.5.1}$$

$$y_s(t) \;=\; C_s x_s(t), \tag{8.5.2}$$

where $x_s(t) \in \mathbb{R}^n$, $u(t) \in \mathbb{R}^1$ and $y_s(t) \in \mathbb{R}^1$ are the measurable state vector, the scalar input and the output, respectively, and $A_s \in \mathbb{R}^{n \times n}$, $B_s \in \mathbb{R}^{n \times 1}$ are the unknown parameter matrices. Moreover, it is assumed that $C_s \in \mathbb{R}^{1 \times n}$ is known and $L_s(t)$ is the unknown varying time-delay. k_0, the upper bound of $\| B_s \|$, is also assumed known.

The purpose of identification is to generate on-line estimates for the unknown matrices A_s, B_s and the unknown varying time delay $L_s(t)$. Therefore the following tracking model is constructed

$$\dot{x}_m(t) \;=\; A_m(t) x_m(t) + B_m(t) u(t - L_m(t)), \tag{8.5.3}$$

$$y_m(t) \;=\; C_s x_m(t), \tag{8.5.4}$$

where $x_m(t) \in \mathbb{R}^n$, $u(t) \in \mathbb{R}^1$ and $y_m(t) \in \mathbb{R}^1$ is the state vector, the scalar input and the output of the tracking model, respectively. $A_m(t) \in \mathbb{R}^{n \times n}$, $B_m(t) \in \mathbb{R}^{n \times 1}$ and $L_m(t) \in \mathbb{R}^1$ are the parameter matrices and the varying time delay of the tracking model, which represent the estimated values of the unknown matrices A_s, B_s and the varying delay $L_s(t)$.

Define the error vector as

$$e(t) = x_s(t) - x_m(t), \tag{8.5.5}$$

then it can be obtained that

$$
\begin{aligned}
\dot{e}(t) \;=\;& \dot{x}_s(t) - \dot{x}_m(t) \\
\;=\;& A_s x_s(t) + B_s u(t - L_s(t)) - A_m(t) x_m(t) - B_m(t) u(t - L_m(t)) \\
\;=\;& A_0 e(t) + (A_s - A_0) x_s(t) - (A_m(t) - A_0) x_m(t) + \\
& B_s u(t - L_s(t)) - B_m(t) u(t - L_m(t)) \\
\;=\;& A_0 e(t) + \theta x_s(t) - \beta x_m(t) + B_s \Delta u(t) + \tilde{B} u(t - L_m(t)), \tag{8.5.6}
\end{aligned}
$$

where A_0 is a stable constant matrix. It is also assumed that there are two matrices, $P = P^T > 0$ and $Q = Q^T > 0$, such that

$$A_0^T P + P A_0 = -Q, \tag{8.5.7}$$

n equation (8.5.6), the following notations are used

$$\theta = (A_s - A_0) \in \mathbb{R}^{n \times n}, \tag{8.5.8}$$

$$\beta = (A_m(t) - A_0) \in \mathbb{R}^{n \times n}, \tag{8.5.9}$$

$$\tilde{B} = B_s - B_m(t) \in \mathbb{R}^{n \times 1}, \tag{8.5.10}$$

$$\Delta u(t) = u(t - L_s(t)) - u(t - L_m(t)). \tag{8.5.11}$$

n order to obtain an adaptive updating laws for $A_m(t)$, $B_m(t)$ and $L_m(t)$ such that

$$\lim_{t \to \infty} e(t) = 0, \tag{8.5.12}$$

he following Lyapunov function is constructed

$$V(A_m, B_m, L_m) = \frac{1}{2}(e^T(t)Pe(t) + tr(\beta^T \Gamma \beta) +$$

$$+ \tilde{B}^T \Lambda \tilde{B} + \sigma L_m^2(t)), \tag{8.5.13}$$

vhere $tr(.)$ denotes the trace of the matrix $(.)$, $\Gamma \in \mathbb{R}^{n \times n}$ and $\Lambda \in \mathbb{R}^{n \times n}$ are positive efinite matrices, and $\sigma \in \mathbb{R}^1$ is a positive constant. The differentiation of $V(A_m, B_m, L_m)$ vith respect to t is therefore given by

$$\dot{V}(A_m, B_m, L_m) = -\frac{1}{2}e^T(t)Qe(t) + e^T(t)P[\theta x_s(t) - \beta x_m(t) + B_s \Delta u(t) +$$

$$+ \tilde{B}u(t - L_m(t))] + tr(\beta^T \Gamma \dot{\beta}) + \tilde{B}^T \Lambda \dot{\tilde{B}} + \sigma L_m(t)\dot{L}_m(t)$$

$$= -\frac{1}{2}e^T(t)Qe(t) + e^T(t)P\theta x_s(t) - e^T(t)P\beta x_m(t) +$$

$$e^T(t)PB_s \Delta u(t) + e^T(t)P\tilde{B}u(t - L_m(t)) + tr(\beta^T \Gamma \dot{\beta}) +$$

$$\tilde{B}^T \Lambda \dot{\tilde{B}} + \sigma L_m(t)\dot{L}_m(t). \tag{8.5.14}$$

,et

$$\dot{A}_m(t) = \Gamma^{-1}Pe(t)x_m^T(t), \tag{8.5.15}$$

$$\dot{B}_m(t) = -\Lambda^{-1}Pe(t)u(t - L_m(t)), \tag{8.5.16}$$

hen $\dot{V}(A_m, B_m, L_m)$ becomes

$$\dot{V}(A_m, B_m, L_m) = -\frac{1}{2}e^T(t)Qe(t) + e^T(t)P\theta x_s(t) +$$

$$+ e^T(t)PB_s \Delta u(t) + \sigma L_m(t)\dot{L}_m(t). \tag{8.5.17}$$

If we can make

$$e^T(t)P\theta x_s(t) + e^T(t)PB_s\Delta u(t) + \sigma L_m(t)\dot{L}_m(t) < 0 \qquad (8.5.18)$$

then

$$\dot{V}(A_m, B_m, L_m) < -\frac{1}{2}e^T(t)Qe(t) < 0 \qquad (8.5.19)$$

where $e(t) \to 0$, as $t \to \infty$. For this purpose, let

$$\sigma L_m(t)\dot{L}_m(t) = -e^T(t)P(\dot{x}_s(t) - A_0 x_s(t)) -$$
$$-k_0 \left\| e^T(t) \right\| \| P \| |u(t - L_m(t))| \qquad (8.5.20)$$

then it can be seen that

$$\sigma L_m(t)\dot{L}_m(t) + e^T(t)P(\theta x_s(t) + B_s\Delta u(t)) =$$
$$-k_0 \left\| e^T(t) \right\| \| P \| |u(t - L_m(t))| - e^T(t)PB_s u(t - L_m(t)) < 0, \qquad (8.5.21)$$

and (8.5.19) holds. The following theorem summaries the results obtained so far.

Theorem 8.1 *If there exists k_0 such that*

$$\| B_s \| < k_0 < +\infty, \qquad (8.5.22)$$

then the parameter estimation process:

$$\begin{cases} \dot{e}(t) = A_0 e(t) + \theta x_s(t) - \beta x_m(t) + B_s\Delta u(t) + \tilde{B}u(t - L_m(t)), \\ \dot{A}_m(t) = \Gamma^{-1}Pe(t)x_m^T(t), \\ \dot{B}_m(t) = -\Lambda^{-1}Pe(t)u(t - L_m(t)), \\ \sigma L_m(t)\dot{L}_m(t) = -e^T(t)P(\dot{x}_s(t) - A_0 x_s(t)) - k_0 \left\| e^T(t) \right\| \| P \| |u(t - L_m(t))|, \end{cases} \qquad (8.5.23)$$

is globally stable and

$$\lim_{t \to \infty} e(t) = 0. \qquad (8.5.24)$$

Although the identification law of the matrices $A_m(t)$ and $B_m(t)$ and the time delay $L_m(t)$ enable the error of the system bounded and $e(t) \to 0$, in practical systems this updating mechanism is difficult to implement since it contain pure differentiation $\dot{x}_s(t)$ which is very sensitive to measurement noises. Therefore, a filter should be included. In this case, we have the following corollary.

Corollary 8.1 *If there exists k_0 such that*

$$\| B_s \| < k_0 < +\infty, \tag{8.5.25}$$

then the parameter estimation process:

$$
\begin{cases}
\dot{e}(t) = A_0 e(t) + \theta x_s(t) - \beta x_m(t) + B_s \Delta u(t) + \tilde{B} u(t - L_m(t)), \\
\dot{A}_m(t) = \Gamma^{-1} P e(t) x_m^T(t), \\
\dot{B}_m(t) = -\Lambda^{-1} P e(t) u(t - L_m(t)), \\
\rho \dot{\xi}(t) = -\xi(t) + \dot{x}_s(t), \xi(t) \in \mathbb{R}^n, \rho > 0 \\
\sigma L_m(t) \dot{L}_m(t) = -(e^T(t) P(\rho \dot{\xi}(t) - A_0 x_s(t)) - \\
\qquad -k_0 \left\| e^T(t) \right\| \| P \| |u(t - L_m(t))| + e^T(t) P \xi(t),
\end{cases} \tag{8.5.26}
$$

is globally stable with

$$\lim_{t \to \infty} e(t) = 0. \tag{8.5.27}$$

Proof: Following the same line as that in Theorem 8.1, it can be obtained that

$$\dot{V}(A_m, B_m, L_m) < -\frac{1}{2} e^T(t) Q e(t) < 0 \tag{8.5.28}$$

$\square\square$

8.5.2 Adaptive pole-placement controller scheme

Having estimated the unknown matrices A_s, B_s and the varying time delay $L_s(t)$, an adaptive control algorithm, which is based on the estimates of $A_m(t)$, $B_m(t)$ and $L_m(t)$, can be constructed. In order to ensure a desirable closed-loop dynamic performance and robustness, pole-placement control strategy shown in Fig. 8.6 is discussed in this subsection.

Since the process model without the time delay (i.e. $L_m(t) = 0$) can be regarded as a predictor, at a fixed time t the transfer function of the process model without the time delay can therefore be expressed as follows

$$C_s(sI - A_m(t))^{-1} B_m(t) =$$

$$\frac{b_1(t)s^{n-1} + b_2(t)s^{n-2} + \ldots\ldots + b_{n-1}(t)s + b_n(t)}{s^n + a_1(t)s^{n-1} + a_2(t)s^{n-2} + \ldots\ldots + a_{n-1}(t)s + a_n(t)} \tag{8.5.29}$$

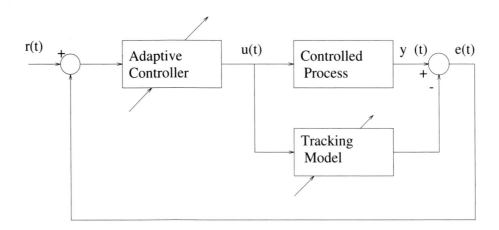

Figure 8.6: The Smith predictor control scheme.

where, according to the Leverrier-Sourian-Faddeeva-Frame formula in [77], $a_i(t)$ and $b_i(t)$ $(i = 1, 2, ..., n)$ are calculated on-line as follows:

$$M_1(t) = I,$$

$$a_1(t) = -tr(A_m(t)),$$

$$M_2(t) = M_1(t)A_m(t) + a_1(t)I,$$

$$a_2(t) = -\frac{1}{2}tr(M_2(t)A_m(t)),$$

$$M_3(t) = M_2(t)A_m(t) + a_2(t)I,$$

$$a_3(t) = -\frac{1}{3}tr(M_3(t)A_m(t)),$$

$$......$$

$$M_{n-1}(t) = M_{n-2}(t)A_m(t) + a_{n-2}(t)I,$$

$$a_{n-1}(t) = -\frac{1}{n-1}tr(M_{n-1}(t)A_m(t)),$$

$$M_n(t) = M_{n-1}(t)A_m(t) + a_{n-1}(t)I,$$

$$a_n(t) = -\frac{1}{n}tr(M_n(t)A_m(t)),$$

$$b_j(t) = C_s M_j(t)B_m(t), \quad (j = 1, 2, ..., n).$$

Based on the transfer function (8.5.29), a predictor in a controllable cannonical form can be given by

$$\dot{x}_p(t) = A_p(t)x_p(t) + B_p(t)u(t) \tag{8.5.30}$$

$$y_p(t) = C_p(t)x_p(t) \tag{8.5.31}$$

where

$$A_p(t) = \begin{bmatrix} -a_1(t) & -a_2(t) & \cdots & -a_{n-1}(t) & -a_n(t) \\ 1 & 0 & \cdots & 0 & 0 \\ 0 & 1 & & 0 & 0 \\ \vdots & & \ddots & & \vdots \\ 0 & 0 & & 1 & 0 \end{bmatrix},$$

$$B_p(t) = [1, 0, \ldots, 0]^T,$$

$$C_p(t) = [b_1(t), b_2(t), \ldots, b_n(t)],$$

and $x_p(t) \in \mathbb{R}^n$, $u(t) \in \mathbb{R}^1$ and $y_p(t) \in \mathbb{R}^1$ are the state vector, the scalar input and the output of the predictor, respectively, whilst $A_p(t) \in \mathbb{R}^{n \times n}$, $B_p(t) \in \mathbb{R}^{n \times 1}$ and $C_p(t) \in \mathbb{R}^{1 \times n}$ are the parameter matrices of the predictor.

When $e(t) \to 0$, the output of the predictor, after being delayed for $L_s(t)$, is that of the controlled plant. As a result, the performance of the predictor represents that of the plant. A controller can then be constructed so that the performance of the closed loop system is desirable. For this purpose, a combination of state feedback and an input feedforward in the predictor is constructed and the controller is given by

$$u(t) = H_p(t)(y_r(t) - \epsilon(t)) - K_p(t)x_p(t), \tag{8.5.32}$$

where $y_r(t) \in \mathbb{R}^1$ is the reference input; $\epsilon(t) = y_s(t) - y_m(t)$ is used to compensate the modelling error and external disturbance. In fact, $\epsilon(t)$ will trend to zero as $t \to \infty$. $H_p(t) \in \mathbb{R}^1$ and $K_p(t) \in \mathbb{R}^{1 \times n}$ are the parameter matrices to be designed. Substituting Eq. (8.5.32) into Eq. (8.5.31), it can be obtained that

$$\dot{x}_p(t) = (A_p(t) - B_p(t)K_p(t))x_p(t) + B_p(t)H_p(t)(y_r(t) - \epsilon(t)). \tag{8.5.33}$$

Let the desired characteristic polynomial of the closed-loop system be

$$T(s) = s^n + d_1 s^{n-1} + \ldots\ldots + d_{n-1}s + d_n, \tag{8.5.34}$$

where d_i $(i = 1, 2, \ldots, n)$ are the constants and express the characteristic polynomial of the predictor as

$$D_o(s) = s^n + a_1(t)s^{n-1} + \ldots\ldots + a_{n-1}(t)s + a_n(t). \tag{8.5.35}$$

By applying pole-placement, it can be seen that

$$K_p(t) = (\bar{d}_o - \bar{a}_o(t)),\tag{8.5.36}$$

where

$$\bar{d}_o = [d_1, d_2, ..., d_n],$$

$$\bar{a}_o(t) = [a_1(t), a_2(t), ..., a_n(t)].$$

Therefore, $K_p(t)$ can be uniquely evaluated. To decide $H_p(t)$, additional requirements on the closed loop system performance should be considered. One of such requirements is zero steady-state response subject to a step reference signal. By using the well known final value theorem in Laplace transform, we have

$$\lim_{s \to 0}[C_p(t)(sI - A_p(t) + B_p(t)K_p(t))^{-1}B_p(t)H_p(t)] = 1\tag{8.5.37}$$

thus the updating law for $H_p(t)$ can be given by

$$H_p(t) = -1/(C_p(t)(A_p(t) - B_p(t)K_p(t))^{-1}B_p(t))\tag{8.5.38}$$

To summarize, the following theorem is obtained.

Theorem 8.2 *When the adaptive control law represented by Eqs. (8.5.15)-(8.5.16), (8.5.20), (8.5.32), (8.5.36) and (8.5.38) are applied to the system (8.5.1) and (8.5.2), the resulting closed-loop system is globally stable and*

$$\lim_{t \to \infty}(r(t) - y_s(t)) = 0.\tag{8.5.39}$$

To illustrate the design procedure, let us consider the following example.

Example 8.2 The plant in this case is a second order dynamic system with time delay. It is described in an observable canonical form as follows.

$$\dot{x}_s(t) = A_s x_s(t) + B_s u(t - L_s(t))\tag{8.5.40}$$

$$y_s(t) = C_s x_s(t)\tag{8.5.41}$$

where

$$A_s = \begin{bmatrix} -a_{s1} & 1 \\ -a_{s2} & 0 \end{bmatrix}\tag{8.5.42}$$

$$B_s = [b_{s1}, b_{s2}]^T\tag{8.5.43}$$

$$C_s = [1, \quad 0]\tag{8.5.44}$$

The form of model is as Eqs. (8.5.3) and (8.5.4), with

$$A_m(t) = \begin{bmatrix} -a_{m1}(t) & 1 \\ -a_{m2}(t) & 0 \end{bmatrix}, \tag{8.5.45}$$

$$B_m(t) = [b_{m1}(t), b_{m2}(t)]^T \tag{8.5.46}$$

The predictor model is therefore given by Eqs. (8.5.30) and (8.5.31) with

$$A_p(t) = \begin{bmatrix} -a_{p1}(t) & -a_{p2}(t) \\ 1 & 0 \end{bmatrix} \tag{8.5.47}$$

$$B_p(t) = [1, \quad 0]^T \tag{8.5.48}$$

$$C_p(t) = [c_{p1}(t), c_{p2}(t)] \tag{8.5.49}$$

Let

$$T(s) = s^2 + 3.5s + 4 \tag{8.5.50}$$

be the desired characteristic polynomial, then the state feedback matrix can be expressed as

$$K_p = [3.5 - a_{p1}(t), 4 - a_{p2}(t)] \tag{8.5.51}$$

and

$$H_p(t) = \frac{4}{c_{p2}(t)} \tag{8.5.52}$$

In addition, we choose

$$A_0 = \begin{bmatrix} -3 & 1 \\ -2 & 0 \end{bmatrix} \tag{8.5.53}$$

$$P = \begin{bmatrix} 2/3 & -1/2 \\ -1/2 & 1 \end{bmatrix} \tag{8.5.54}$$

The estimation of the parameters, $a_{m1}(t)$, $a_{m2}(t)$, $b_{m1}(t)$, $b_{m2}(t)$ and $L_m(t)$, in the process model can be then obtained directly from Eqs. (8.5.15), (8.5.16) and (8.5.20). The sampling period for this particular example is set to 0.1 second and the reference input, $y_r(t)$, is again a square wave of period 12 seconds, as is shown in Figs. 8.7 and 8.8.

Figures 8.7 shows the responses of the system using the adaptive pole-placement controller, where $y_s(t)$ is the output response of the plant, $y_m(t)$ is the output of the process model and $\epsilon(t)$ is the output error between the plant and the estimated model. Figure 8.8 presents the performances of the system using the standard Smith predictor control (i.e. the estimation of the parameters are disconnected).

In Figs 8.7 and 8.8 the parameters and the time delay of the plant are same as those of the process model from 0s to 12s, i.e. $a_{s1} = a_{m1}(0) = 2, a_{s2} = a_{m2}(0) = 6.67545, b_{s1} =$

$b_{m1}(0) = 1, b_{s22} = b_m(0) = 3.16228$ and $L_s(t) = L_m(0) = 0.98$. Then they are changed twice during the simulation phase. The first change is made at $t = 12s$, where new set of parameter values, $a_{s1} = 1$, $a_{s2} = 4.67545$, $b_{s1} = 0.8$, $b_{s2} = 2.53$ and $L_s(t) = 0.77$, are used. The second change is created when $t = 36s$ with new values being set to $a_{s1} = 0.9$, $a_{s2} = 3.68$, $b_{s1} = 0.7$, $b_{s2} = 2.2136$ and $L_s(t) = 0.66$.

It can be seen from the simulation results that the adaptive pole-placement controller can overcome the difficulties caused by the time delay, and thus produces a better dynamic response than that of the standard Smith predictor control, which displays sensitive to the variations of process parameters and the time delay.

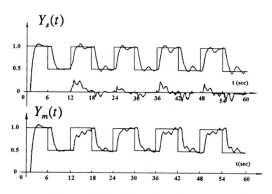

Figure 8.7: The adaptive control.

8.6   Conclusions

It is well known that time delay, especially variable time delay in industrial processes, is a key factor which affects the performance of the closed loop systems. The accurate estimation is therefore very important to the design and tuning phases of controllers. In this chapter, we have discussed an adaptive control strategy for unknown linear systems with variable time delay. In specific, a model reference adaptive predictor control strategy, which is partly based upon the structure of the Smith predictor control, is described.

Two different methods have been used to derive the adaptive updating laws for the estimation of the variable time delay. The first method is simply the application of parametric optimization theory, whilst the second method is based on the use of Lyapunov stability theory which ensures the stability of the closed loop system. However, the second method is more complicated than the first one in the sense that the state vector is required

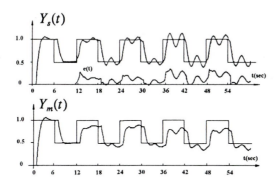

Figure 8.8: The Smith predictor control.

to estimate the variable time delay. Two examples are included to illustrate the design procedure discussed in this chapter and good results are obtained in each cases.

Chapter 9

Rule-based Adaptive Control Systems Design

9.1 Introduction

Over the last decade, many methods have been developed in the research areas of intelligent and expert control systems. In general, these systems consist of the plant, a controller and a higher level knowledge base system which is used to select the structure of the controller and tune its parameters in order to achieve the required performance for the complete system. The early developments can be traced back to the tuning of a PI controller proposed by Ziegler and Nichols in 1942 [183], where two methods, the reaction curve and continuous cycling, were developed. With the first method, controller tuning is carried out by using an open loop step response of the process and the tuning is then based on the effective time delay and the maximum slope of the response curve. In their second method, the tuning is achieved within a closed loop framework where a gradually increasing proportional gain is used. The gain is increased until an oscillation with constant amplitude occurs in the output and the frequency and magnitude of this oscillation are used to set the controller parameters. Improvements to the Ziegler-Nichols tuning methods have been made by Cohen and Coon [32], Fertik [49] and Astrom and Hagglund [10]. The method of Astrom and Hagglund [10], essentially automates and improves the Ziegler-Nichols continuous cycling method. In their method, a low-gain PI controller is initially used to bring the process near the set point and then a relay replaces the PI controller so as to reduce the size of the oscillation. With proper adjustment of the relay output, a reasonable amplitude for the oscillations is obtained and again, the frequency and the amplitude are used to tune the controller. In these approaches, the tuning is achieved by directly using information contained in the output data of the system (the maximum slope of the response curve, the effective pure time delay, peak overshoot and rise time, for example). This leads to a straightforward method of tuning the gains for simple control schemes, such as PID controllers. However, in almost all of the methods

the final tuning of the controllers is manual. This is due to the fact that the time domain behaviour of modern processes cannot be related exactly to the design parameters
of the controller tuning methods. Moreover, these methods can be difficult to apply to
complex processes such as those with high dynamic order or with time-varying characteristics. More recently, autonomous online methods based upon neurofuzzy algorithms have
been developed[85] which utilise feature or label vectors that depict closed loop behaviour
(eg. rise time, peak overshoot and steady state error, etc) to synthetise the gains of a
PID controller via associative memory mappings (see chapter 3). This approach applies
to high order, nonlinear, time varying and time delayed plants with unknown dynamics
and is based on the assumption that a PID controller exists to provide adquate control
performance.

The expert system tuner developed by Jones and Porter [72] is capable of tuning a
wide variety of plants, but different rules are needed for different types of plants. Based on
a conceptual model of the plant, different rules are selected to tune the controller. In their
method, the expert tuner continually checks this conceptual model as the system is tuned
and a consistency enforcer is used to change the conceptual model if the transient response
does not match the predicted response. Consequently, the structure of this expert system
tuner is fairly complex as it involves two knowledge bases (a consistency enforcer and the
use of a solution blackboard) and the number of rules is relatively high.

Therefore, in order to overcome the difficulties discussed in the previous paragraphs
and to obtain the high performance and easily implemented controller for the real-time
control of practical industrial systems, Wang and Daley [151] developed a simple but generally applicable rule based system for the controller's tuning. In their method, the rule
based identification methods, which was established by Daley and Wang in 1990 [38] and
Wang and Jones in 1991 [149], is used to obtain a linear model between the tracking error
and the set point. Using this model and the existing controller settings, the characteristic
of the unknown plant can be formulated. As a result, new setting for both the structure
and the parameters can be obtained. In this chapter, rule-based identification and adaptive control tuning strategy will be discussed, which can be regarded as an alternative
approach to the design of adaptive control systems for either linear or nonlinear unknown
systems.

9.2 Systems Representation

Let the unknown discrete time system be expressed in a general form as

$$S(d, n, m, p) : (u(k), w(k)) \rightarrow y(k) \tag{9.2.1}$$

where S is referred as the model or map of the system, including information about the
system time delay, d, the structure order n and m, and the unknown parameter vector p.
$u(k)$ is the measured input to the system; $w(k)$ is the noise signal, which may be either

random or an undefined deterministic process and $y(k)$ is the output of the system, which can be expressed as the composite response to the actual signal a and to that of the noise ω

$$y(k) = y_a(k) + y_\omega(k) \tag{9.2.2}$$

It is further assumed in Wang and Jones [149] that system S is stable in the sense that

$$\|S(d, n, m, p)\| \leq c_0 < \infty \tag{9.2.3}$$

It can be seen that system Eq. (9.2.1) includes the ARMAX model

$$A(q^{-1})y(k) = B(q^{-1})u(k - d) + C(q^{-1})d(k) \tag{9.2.4}$$

and the more general NARMAX model

$$\begin{aligned}
y(k) &= f(y(k-1), y(k-2), \cdots, y(k-n), u(k-d), u(k-d-1), \cdots, \\
&\quad u(k-d-m), \omega(k), \omega(k-1), ..., \omega(k-n))
\end{aligned} \tag{9.2.5}$$

For example, in the former case, the coefficients in polynomials $A(q^{-1})$, $B(q^{-1})$ and $C(q^{-1})$ constitute the unknown parameter vector p, whilst the orders of these polynomials determine the structure order of the system.

The aim of rule based identification is therefore to establish a set of rules which will be used to find a matching model $M(\hat{d}, \hat{n}, \hat{m}, \hat{p})$ such that the output of $M(\hat{d}, \hat{n}, \hat{m}, \hat{p})$ should follow the signal $y(k)$ to a specified accuracy, when the same step input signal is applied. It can be seen that the matching model can be either linear or nonlinear, and it may not necessarily be in the exact form as the original model!

The form of the matching model considered by Wang and Jones [149] is a linear and noise free model, the advantage of using such a linear matching model rather than a nonlinear matching model is that the tuning can be easily realized through its poles and zeros. The matching model takes the form of

$$M(\hat{d}, \hat{n}, \hat{m}, \hat{p}) : (u(k), 0) \to y_m(k) \tag{9.2.6}$$

and is assumed to be an ARMA model with time delay $\hat{d}$, structure orders $\hat{n}$ and $\hat{m}$ and a pole and zero set $\hat{p}$. $u(k)$ is the input signal and $y_m(k)$ is the output signal of the matching model, both in discrete time. More specifically, the aim of the modelling is to find a set of rules that can be used to find a noise-free ARMA model approximation of the unknown system S. The idea is to tune the poles and zeros of the matching model such that the difference between the response of the system and that of the matching model is made as small as possible. That is, if the response of the matching model is faster than that of the actual system, then the magnitude of some of the poles of the matching model should be increased, otherwise, the magnitude of some poles should be reduced.

Because the rules to be considered are for adjusting the poles and zeros of the matching model $M(\hat{d}, \hat{n}, \hat{m}, \hat{p})$ such that our purpose is achieved, in the following we will refer to the matching model as a pole-zero model.

Let the matching model $M(\hat{d}, \hat{n}, \hat{m}, \hat{p})$ be of the form

$$M(\hat{d}, \hat{n}, \hat{m}, \hat{p}) = \frac{\hat{b}_0 q^{-\hat{d}} \Pi_{j=1}^{\hat{m}}(1 - b_j q^{-1})}{\Pi_{i=1}^{\hat{n}}(1 - a_i q^{-1})} \qquad (9.2.7)$$

$$y_m(k) = M(\hat{d}, \hat{n}, \hat{m}, \hat{p}) u(k) \qquad (9.2.8)$$

With this model, the following state space model realization can be readily obtained.

$$x(k) = [I - E_1]^{-1} A_1 x(k - 1) + [I - E_1]^{-1} \bar{b}_1 u(k) \qquad (9.2.9)$$
$$v(k) = c_1 x(k) \qquad (9.2.10)$$
$$w(k) = [I - E_2]^{-1} B_1 w(k - 1) + [I - E_2]^{-1} \bar{b}_2 b_0 v(k - d) \qquad (9.2.11)$$
$$y_m(k) = c_2 w(k) \qquad (9.2.12)$$

with

$$A_1 = \begin{bmatrix} a_1 & 0 & 0 & . & . & 0 \\ 0 & a_2 & 0 & . & . & 0 \\ . & . & . & . & . & . \\ . & . & . & . & . & . \\ . & . & . & . & . & . \\ 0 & 0 & . & . & . & a_{\hat{n}} \end{bmatrix} \in R^{\hat{n} \times \hat{n}} \qquad (9.2.13)$$

$$\bar{b}_1 = \begin{bmatrix} 1 & 0 & 0 & . & . & 0 \end{bmatrix}^T \in R^{\hat{n}} \qquad (9.2.14)$$

$$c_1 = \begin{bmatrix} 0 & 0 & 0 & . & . & 1 \end{bmatrix} \in R^{1 \times \hat{n}} \qquad (9.2.15)$$

$$E_1 = \begin{bmatrix} 0 & 0 \\ I_{\hat{n}-1} & 0 \end{bmatrix} \in R^{\hat{n} \times \hat{n}}; \quad B_1 = \begin{bmatrix} 0 & 0 \\ Z_{\hat{m}} & 0 \end{bmatrix} \in R^{(\hat{m}+1) \times (\hat{m}+1)} \qquad (9.2.16)$$

$$Z_{\hat{m}} = diag(-b_1 \quad -b_2 \quad . \quad . \quad . \quad -b_{\hat{m}}) \in R^{\hat{m} \times \hat{m}} \qquad (9.2.17)$$

$$\bar{b}_2 = \begin{bmatrix} 1 & 0 & 0 & . & . & 0 \end{bmatrix}^T \in R^{(\hat{m}+1) \times 1} \qquad (9.2.18)$$

$$c_2 = \begin{bmatrix} 0 & 0 & . & . & 0 & 1 \end{bmatrix} \in R^{1 \times (\hat{m}+1)} \qquad (9.2.19)$$

$$E_m = \begin{bmatrix} 0 & 0 \\ I_{\hat{m}} & 0 \end{bmatrix} \qquad (9.2.20)$$

and $I_{\hat{n}-1}, I_{\hat{m}}$ are unit matrices of dimensions $\hat{n} - 1$ and $\hat{m}$, respectively.

However, if $y(k)$ is oscillatory, clearly the matching model $M(\hat{d}, \hat{n}, \hat{m}, \hat{p})$ should include some complex poles and zeros. In this case, the matching model is assumed to take the form

$$M(\hat{d}, \hat{n}, \hat{m}, \hat{p}) = \frac{b_0 q^{-\hat{d}} \Pi_{j=1}^{\hat{m}_1}(1 - b_j q^{-1}) \Pi_{j=1}^{\hat{m}_2}(1 - 2\gamma_j q^{-1} + \nu_j q^{-2})}{\Pi_{i=1}^{\hat{n}_1}(1 - a_i q^{-1}) \Pi_{j=1}^{\hat{n}_2}(1 - 2\alpha_j q^{-1} + \omega_j q^{-2})} \qquad (9.2.21)$$

where

$$\nu_i = \gamma_i^2 + \theta_i^2 \qquad (9.2.22)$$

$$w_j = \alpha_j^2 + \beta_j^2 \tag{9.2.23}$$
$$\theta_i \neq 0; \quad \beta_j \neq 0 \tag{9.2.24}$$
$$\hat{n}_1 + \hat{n}_2 = \hat{n} \tag{9.2.25}$$
$$\hat{m}_1 + \hat{m}_2 = \hat{m} \tag{9.2.26}$$

By a similar procedure, a state realization model can be obtained in pole-zero form.

9.3   Meta Identification Rules

The knowledge encoding within the system identifier takes the form of a set of rules. They are used to give an estimation of time delay, to adjust the poles and zeros of the matching model $M(\hat{d}, \hat{n}, \hat{m}, \hat{p})$ to give the best approximation to the original unknown system S, and to determine the minimum orders of the structure of the unknown system S. Two different types of rules are used and they are:

i) meta identification rules;

ii) model adjusting rules.

The meta identification rules contain information about how to use the model adjusting rules. In other words, the meta identification rules are 'fired' by the data obtained from the unit step response test. In this case, two subsets of rules are developed, one, filtering rules; and two, classification rules. The later subset gives information on the type of response and the initial structure of the matching model, such as the initial structure orders and whether the model specified by Eqs. (9.2.7) to (9.2.8) or Eq. (9.2.21) will be used.

9.3.1   Filtering rules

Because the open-loop step response data of an unknown system S is normally corrupted by noise (the type of which is unknown), a set of filtering rules should be constructed in order to eliminate the effect of external noise on the modelling of the system. These rules include the determination of static state noise bounds, the time delay of the system, an up function and a low function of the original data defined below. An illustration of the data from an open-loop step response is shown in Fig. 9.1.

Using prior knowledge of the original system, the regulation time t_s can be determined. Assume that the number of samples after time t_s is N, then the static state bound of the noise can be defined as

$$\epsilon = \frac{\sum_i \bar{y}(k_i)}{N - K} - \frac{\sum_i \underline{y}(k_i)}{K} \tag{9.3.1}$$

where K is the number of local minimum points of the original data after time t_s, and $N - K$ is the number of local maximum points. $\bar{y}(k_i)$ is the local maximum value of the

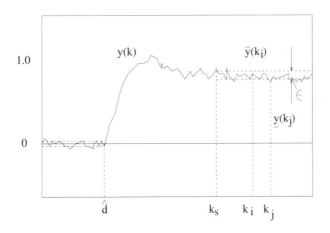

Figure 9.1: A open loop step response.

original response function at time k_i, whereas $\underline{y}(k_i)$ is the local minimum value at time k_i. With ϵ, the time delay $\hat{d}$ can be determined from

$$\hat{d} = min\{k|\,|y(k)| > \epsilon\} \tag{9.3.2}$$

The up function $y_U(k)$ is defined as the direct line connection of the local maximum points of the original system response, and the low function $y_L(k)$ is defined as the direct line connection of the local minimum points. Both functions are shown in Fig. 9.2.

Using these two functions, the first order filtered function, $y^{(1)}(k)$, is defined as the mean

$$y^{(1)}(k) = \frac{1}{2}(y_U(k) + y_L(k)) \tag{9.3.3}$$

The first order filtered function is expected to be much smoother than the original system response, however, it may still contain significant noise. Further filtering is therefore required in some cases. This can be achieved by using the above procedure recursively, until the noise level is within a pre-specified range. For example, the second order filtered function, $y^{(2)}(k)$, can be obtained by averaging the up and low functions for the first order filtered function $y^{(1)}(k)$. Denote $y_U^{(i)}(k)$ and $y_L^{(i)}(k)$ as the up and low functions of the ith order filtered function, respectively, then the $(i+1)$th order filtered function, $y^{(i+1)}(k)$, is given by

$$y^{(i+1)}(k) = \frac{1}{2}(y_U^{(i)}(k) + y_L^{(i)}(k)) \tag{9.3.4}$$

Although the filtering can be stoped at any order, a smooth **noise-free** curve can always be obtained whenever the up function and low function are equal. In either case, the filtered function is denoted as $y_I(k)$ and is used to decide upon the type of response and the initial structure of the matching model.

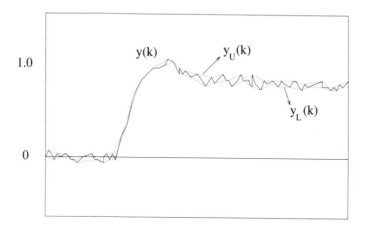

Figure 9.2: Up and low functions.

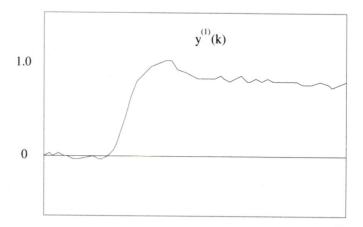

Figure 9.3: The first order filtered function.

9.3.2 Meta identification rules for response type determination

Using function $y_I(k)$, the system response can be characterised by the four disjoint definitions:

i) minimum phase monotone (MPM);

ii) nonminimum phase monotone (NPM);

iii) minimum phase oscillatory (MPO);

iv) nonminimum phase oscillatory (NPO).

To decide the type of the response, four meta rules have been established, which are illustrated as follows

Rule 1:

If $\max_{0<k<t_s} y_I(k)$ is less than or equal to $\frac{1}{N} \sum_{k_i \leq t_s} y_I(k_i)$ and $y_I(k) > y(0)$ for all k, then the type of the system is MPM, and the initial poles and zeros of the matching model are real and located inside the interval $(0, 1)$. The model can therefore be expressed as

$$M_0(\hat{d}, \hat{n}, \hat{m}, \hat{p}) = \frac{b_0 q^{-\hat{d}}}{(1 - a_1 q^{-1})} \tag{9.3.5}$$

Rule 2:

If $\max_{0<k<t_s} y_I(k)$ is less than or equal to $\frac{1}{N} \sum_{k_i \leq t_s} y_I(k_i)$ and $y_I(k) < y(0)$ for some $k < t_s$, then the type of the system is NPM, and the initial poles and zeros of the matching model are real and located inside the interval $(0, 1)$. There always be at least one zero of the system in the interval of $(-\infty, -1)$. For example, the initial matching model can be as follows

$$M_0(\hat{d}, \hat{n}, \hat{m}, \hat{p}) = \frac{b_0 q^{-\hat{d}}(1 + b_1 q^{-1})}{(1 - a_1 q^{-1})(1 - a_2 q^{-1})} \tag{9.3.6}$$

Rule 3:

If $\max_{0<k<t_s} y_I(k)$ is larger than $\frac{1}{N} \sum_{k_i \leq t_s} y_I(k_i)$ and $y_I(k) > y(0)$ for all k, then the type of the system is MPO, and the initial poles and zeros of the matching model are complex and all the zeros, if incorperated, are inside the interval $(0, 1)$. The matching model in this case can be expressed as

$$M_0(\hat{d}, \hat{n}, \hat{m}, \hat{p}) = \frac{b_0 q^{\hat{d}}}{(1 - 2\alpha_j q^{-1} + w_j q^{-2})} \tag{9.3.7}$$

Rule 4:

If $\max_{0<k<t_s} y_I(k)$ is larger than $\frac{1}{N}\sum_{k_i \le t_s} y_I(k_i)$ and $y_I(k) < y(0)$ for some $k < t_s$, then the type of the system is NPO, and some of the initial poles are complex, whilst the number of the zeros is odd and some zeros lie inside the interval $(-\infty, -1)$. The minimum structured model can therefore be expressed as

$$M_0(\hat{d}, \hat{n}, \hat{m}, \hat{p}) = \frac{b_0 q^{\hat{d}}(1 - b_j q^{-1})}{(1 - 2\alpha_j q^{-1} + \omega_j q^{-2})} \qquad (9.3.8)$$

The typical responses for these systems are shown in Figs.9.4-9.7.

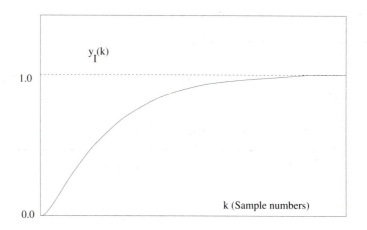

Figure 9.4: Minimum phase monotone response.

9.4 Model Adjusting Rules

The model adjusting rules are used to adjust the poles and zeros of the matching model. The aim is to give a desirable approximation to the output of the open-loop step response of the original system S. The rules for adjusting real poles and zeros are different from those for adjusting complex poles and zeros. The rules are fired by the error between the output of the original system and the output of the matching model. This error can be defined as

$$e(k) = y_m(k) - y(k) \qquad (9.4.1)$$

With this definition, two suberror functions, $e^+(k)$ and $e^-(k)$, are constructed as follows

$$e^+(k) = \begin{cases} 1 + e^+(k-1) & e(k) \ge 0 \\ e^+(k-1) & e(k) < 0 \end{cases} \qquad (9.4.2)$$

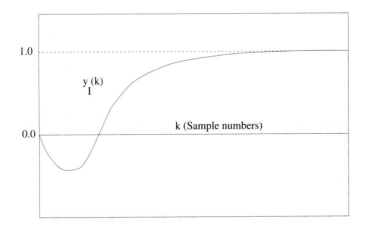

Figure 9.5: Non-minimum phase monotone response.

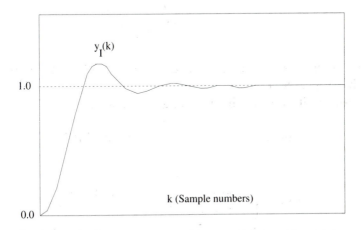

Figure 9.6: Minimum phase oscillatory response.

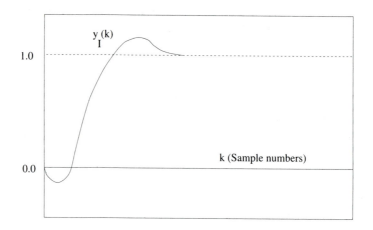

Figure 9.7: Non-minimum phase oscillatory response.

and

$$e^-(k) = \begin{cases} 1 + e^-(k-1) & e(k) < 0 \\ e^-(k-1) & e(k) \geq 0 \end{cases} \tag{9.4.3}$$

The performance function J which characterizes the modelling result and the static state gain $G(0)$ of the original system are defined by

$$J = \frac{\sum_{k_i} |e(k_i)|}{k_i + N} \tag{9.4.4}$$

$$G(0) = \frac{\sum_{k_i} y_I(k_i)}{N} \tag{9.4.5}$$

The rules for adjusting the real poles of the matching model are given as follows:

Model adjusting rule 1

If $J > c_0$, then the ith pole of the matching model is tuned by

$$a_i(k) = a_i(k-1) - f_0(e^+(k) - e^-(k))J \tag{9.4.6}$$

otherwise

$$a_i(k) = \begin{cases} 1.002a_i(k-1) & (e^+ - e^-) \geq 4 \\ 0.998a_i(k-1) & (e^+ - e^-) < -2 \end{cases} \tag{9.4.7}$$

where c_0 and f_0 are positive small constants. The zeros of the matching model are adjusted so that the static state gain of the matching model, $G_m(0)$, is made equal to that of the original system, i.e.

$$b_0 \Pi_{j=1}^{\hat{m}}(1 - b_j(k)) = G(0) \Pi_{i=1}^{\hat{n}}(1 - a_i(k)) \tag{9.4.8}$$

The adjustment of the complex poles is similar to that for real poles. However, the real part and imaginary parts are adjusted separately.

Model adjusting rule 2:

Let the real and the imaginary part of a pole p_i be

$$p_i(k) = \alpha_i(k) + j\beta_i(k) \tag{9.4.9}$$

If $J > c_0$, then

$$\begin{aligned}
\alpha_i(k) &= \alpha_i(k-1) - f_0 J(e^+(k) - e^-(k)) - \\
&\quad - f_1(|e^+(k)| + |e^-(k)|) \tag{9.4.10} \\
\beta_j(k) &= \beta_j(k-1) - f_0 J(e^+(k) - e^-(k)) - \\
&\quad - f_1(|e^+(k)| + |e^-(k)|) \tag{9.4.11}
\end{aligned}$$

otherwise

$$\alpha_i(k) = \begin{cases} 1.002\alpha_i(k-1) & (e^+ - e^-) \geq 4 \\ 0.998\alpha_i(k-1) & (e^+ - e^-) < -2 \end{cases} \tag{9.4.12}$$

$$\beta_i(k) = \begin{cases} 1.002\beta_i(k-1) & (e^+ - e^-) \geq 4 \\ 0.998\beta_i(k-1) & (e^+ - e^-) < -2 \end{cases} \tag{9.4.13}$$

where f_1 is also a small positive number. The zeros of the matching model are also adjusted by Eq. (9.4.8).

The key point here is that the model should be tuned such that the two suberror vectors are close to each other. For the tuning of the real poles, the principle of the tuning is to increase the magnitude of some poles if the response of the matching model is faster than that of the original system. Otherwise, the magnitude of some poles should be reduced. By combining the meta identification rules and model adjusting rules, the main algorithm for this rule based identifier can be established as follows.

Main algorithm: Part I

i) Determine the noise bound ϵ and estimate the time delay $\hat{d}$;

ii) Use filtering rules to generate $y_I(k)$;

iii) Apply meta identification rules 1-4 to decide upon the type of system response and to choose the initial structure of the matching model, with initial structure orders being $\hat{n}(0)$ and $\hat{m}(0)$;

iv) Calculate the step response of the matching model to obtain $y_m(k)$, and use the model adjusting rules to tune the poles and zeros of the matching model. Repeat this stage until

$$\left|(e^+(k) - e^-(k))\right| \le 4; \tag{9.4.14}$$

v) Evaluate the corresponding performance function $J(\hat{n}(0), \hat{m}(0))$.

It can be seen that in the above algorithm, the initial selected structure orders are fixed. This may not normally lead to a minimum structured estimate, and indeed, the estimate may not even converge. In the following, an additional algorithm is constructed in order to generate the estimate for the minimum structure orders of the system. The idea is that after the convergence of the above algorithm, the value of $J(\hat{n}(0), \hat{m}(0))$ is checked against the targeted value, and the structure orders are increase by one if, as a result of comparison, the actual $J(\hat{n}(0), \hat{m}(0))$ is not close to its targeted value. The details are given below.

Main algorithm: Part II

i) Increase structure orders $\hat{n}(0)$ and $\hat{m}(0)$ to get the new estimation of the structure orders $\hat{n}(1)$ and $\hat{m}(1)$ as

$$\hat{n}(1) = \hat{n}(0) + 1 \tag{9.4.15}$$
$$\hat{m}(1) = \hat{m}(0) + 1 \tag{9.4.16}$$

and repeat stage $v)$ in Part I to find out $J(\hat{n}(1), \hat{m}(1))$.

ii) Repeat stage $i)$ to find out $J(\hat{n}(i), \hat{m}(i))$, where

$$\hat{n}(i) = \hat{n}(i-1) + 1 \tag{9.4.17}$$
$$\hat{m}(i) = \hat{m}(i-1) + 1 \tag{9.4.18}$$

until there is an i_0 such that

$$|J(\hat{n}(i_0 + 1), \hat{m}(i_0 + 1)) - J(\hat{n}(i_0), \hat{m}(i_0))| \le \epsilon_1 \tag{9.4.19}$$

where ϵ_1 is a pre-specified small number. Then the minimum structure orders of the system are $\hat{n}(i_0 + 1)$ and $\hat{m}(i_0 + 1)$.

As a result, the adjusted poles and zeros, together with the estimated time delay $\hat{d}$ and the structure orders, give an estimation of the original system.

It can be seen in practice that the algorithm will need a number of iterations to converge to a good model. However, the number of iterations will mainly depend on the choice of the initial values of the parameters and the structure orders as well as the signal-to-noise ratio of the step response data. Moreover, although the rule based identification described previously is essentially an off-line method, in practice there are many systems where a step input may be applied during nominal operation. In this circumenstance, the above algorithm may also be used for on-line identification. The control of the paper machine basis weight can be regarded as one such example, where a bump test can be applied to the slice lip opening under the headbox. The magnitude of the input step test signal must be within the safety range, but ideally, as large as possible to excite the system dynamics.

As discussed in [58, 140], there are a number of step response testing techniques described in the literature on network synthesis. They are actually aimed at finding a set of exponentials whose sum approximates the output response as closely as possible. These methods require knowledge of the structure orders of the system and a higher signal-to-noise ratio for the output response data. Therefore they are not as robust as the methods described here. In recent years, graphic based approaches have been developed, due largely to the improvement in computer graphics interface and memory. These methods often require experienced engineers to manipulate the keyboard in order to achieve an accurate model. Rule based identification is more straight forward than the graphical methods simply because the model can be derived completely by a numerical algorithm via an appropriate process data base.

9.5   Controller Tuning via Rule Based Method

9.5.1   System representation

The system considered in this section is shown in Fig .9.8 which consists of three parts:

i) The unknown plant to be controlled which is represented by the transfer function $G(q^{-1})$ relating the input $u(k)$ and the output $y(k)$ as follows;

$$y(k) = G(q^{-1})u(k) \qquad (9.5.1)$$

ii) The controller which is represented by $\hat{C}_i(q^{-1})$ relating the tracking error to the control signal applied to the plant in the following way;

$$u(s) = \hat{C}_i(q^{-1})e(k) \qquad (9.5.2)$$

where

$$e(k) = y_r(k) - y(k) \qquad (9.5.3)$$

is defined as the tracking error, $y_r(k)$ is the demand signal and the subscript i for the controller stands for the tuned controller after ith iteration.

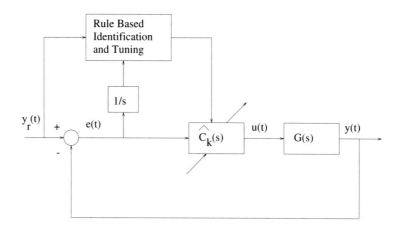

Figure 9.8: The rule based tuning system.

iii) A rule based tuner which consists of the rule based identification algorithm discussed
in sections 9.2-9.4 and a controller calculation unit.

Let the demand input $y_r(k)$ be of the form shown in Fig. 9.9 in which k is an integer
number and T_0 is the period of the demand input, then as will be seen in the next section
that the tuning procedure consists of on-line data collection sub-procedure and controller
tuning sub-procedure. These two sub-procedures operates alternative, i.e., the data of
the integral of the tracking error is collected during the period when $y_r(k) = 1.0$ and the
controller is tunned when $y_r(k) = 0$.

For this structure of systems, the tuning of the controller and the rule-based identification operate in the following sequence.

9.5.2   Controller tuning

The aim of the rule based identification and tuning block in Fig. 9.8 is to tune the
controller $\hat{C}_i(q^{-1})$ in both structure and parameters such that the output of the closed
loop system is stable, satisfies certain requirements for dynamic performance and tracks
the step input $y_r(k)$ without steady state error, i.e.,

$$Lim_{k\to\infty} e(k) = 0 \tag{9.5.4}$$

Denote

$$F_i(n, m, q^{-1}, p) = \frac{e(k)}{(1 - q^{-1})r(k)} = \frac{b_0 \Pi_{j=1}^m (1 - b_j q^{-1})}{\Pi_{i=1}^m (1 - a_i q^{-1})} \tag{9.5.5}$$

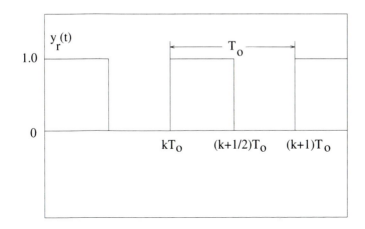

Figure 9.9: The demand signal $r(k)$.

as the $y_r(k)$ to the integral of $e(k)$ transfer function during $[iT, (i+1)T]$, where n and m are structural orders and

$$p = [a_1, \quad a_2, \quad \cdots \quad , a_n, \quad b_1, \quad b_2, \quad \cdots, \quad b_m] \tag{9.5.6}$$

is a parameter vector. With Eq. (9.5.5) and from Fig. 9.8, it can be obtained that

$$F_i(n, m, q^{-1}, p) = \frac{1}{(1 - q^{-1})(1 + \hat{C}_i(q^{-1})G(q^{-1}))} \tag{9.5.7}$$

Let $\hat{F}_i(\hat{n}, \hat{m}, q^{-1}, \hat{p})$ be the estimate of $F_i(n, m, q^{-1}, p)$ and $\hat{n}$, $\hat{m}$ and $\hat{p}$ be estimates of n, m and p, respectively, then the proposed algorithm is given as follows:

i) Based on prior knowledge of the plant, choose the initial controller $\hat{C}_0(q^{-1})$ which stabilizes the closed loop system in Fig .9.8 and let $i = 0$. In the case that the plant is totally unknown, use the rule based identification algorithm in sections 9.2-9.4 to estimate the system first and then choose the initial controller $\hat{C}_0(q^{-1})$ to stabilize the plant;

ii) For $i=1, 2, \ldots$, obtain the measurements of $e(k)$ and $y_r(k)$ and form $\sum_{j=0}^{k} e(j)$ during the period $[iT_0, iT_0 + \frac{1}{2}T_0 i]$

iii) Based on these measurements, the rule based identification algorithm is used to obtain the transfer function $\hat{F}_i(\hat{n}, \hat{m}, q^{-1}, \hat{p})$. Substituting $F_i(n, m, q^{-1}, p)$ with $\hat{F}_i(\hat{n}, \hat{m}, q^{-1}, \hat{p})$ in Eq. (9.5.7) gives

$$\hat{F}_i(\hat{n}, \hat{m}, q^{-1}, \hat{p}) = \frac{1}{(1 - q^{-1})(1 + \hat{C}_{i-1}(q^{-1})G(q^{-1}))} \tag{9.5.8}$$

Then the estimate of the plant $\hat{G}_i(q^{-1})$ is calculated from Eq. (9.5.8) as follows

$$\hat{G}_i(q^{-1}) = \frac{1 - (1 - q^{-1})\hat{F}_i(\hat{n}, \hat{m}, q^{-1}, \hat{p})}{(1 - q^{-1})\hat{C}_{i-1}(q^{-1})\hat{F}_i(\hat{n}, \hat{m}, q^{-1}, \hat{p})} \qquad (9.5.9)$$

iv) During the period $[iT_0, T_0 i + \frac{1}{2}T_0 i]$, if the $\sum_{j=0}^{k} e(j)$ does not converge to a constant value, the structure of the controller is modified by

$$\hat{C}_i(q^{-1}) = \frac{1}{(1 - q^{-1})}\left(\frac{c_{i_0} + c_{i_1}q^{-1}}{c_{i_2} + c_{i_3}q^{-1}}\right)\hat{C}_{i-1}(q^{-1}) \qquad (9.5.10)$$

Otherwise the structure of the controller will not be changed. In the above equation, $c_{i_0}, c_{i_1}, c_{i_2}$ and c_{i_3} are to be determined together with other remaining parameters of the controller in step iv). As a result, this procedure will lead to a fixed structure controller since only the tracking with respect to a step input is considered. For example, if the initial controller is set proportional and there is no integration effect in the forward path of the closed-loop system, then the structure of the controller after the first iteration will be modified to PI control structure.

v) With the estimate $\hat{G}_i(q^{-1})$ and after the necessary structure modification for the controller, determine the parameters of the controller $\hat{C}_i(q^{-1})$ which will result in a better dynamic performance for the closed loop system.

Many methods are now available to design the controller based on the estimate of an unknown plant. In this chapter, a pole-assignment method will be used simply because the requirement of the dynamic performance for the closed loop system can be easily realized via the positioning of desired closed loop poles. Let the desired location of the closed loop poles be

$$(\lambda_1, \lambda_2, ..., \lambda_{\bar{n}}) \qquad (9.5.11)$$

and then the corresponding characteristic polynomial becomes

$$T(q^{-1}) = \Pi_{i=1}^{\bar{n}}(1 - \lambda_i q^{-1}) \qquad (9.5.12)$$

Since the structure order of the controller may be increased during the tuning procedure, it seems, from Eqs (9.5.11) and (9.5.12), that the total number of the desired poles $\bar{n}$ need to be changed as well. This will cause difficulty and increase the complexity of the tuning since some new desired poles for the closed loop system need to be added to form the new set of poles. To avoid this, the number of dominant poles is fixed during the tuning phase, and only non-dominant stable poles are added after the control structure changes. For instance, one can fix $\bar{n}_0$ dominant poles as

$$(\lambda_1, \lambda_2, ..., \lambda_{\bar{n}_0}) \qquad (9.5.13)$$

and if the structure order of the controller is increased by 1, a non-dominant stable pole $\lambda_{\bar{n}_0+1}$ satisfying

$$|\lambda_{\bar{n}_0+1}| = 0.10 \min_{1 \le i \le \bar{n}_0} |\lambda_i| \qquad (9.5.14)$$

can be added to make the total set of the desired poles become

$$(\lambda_1, \lambda_2, ..., \lambda_{\bar{n}_0}, \lambda_{\bar{n}_0+1}) \qquad (9.5.15)$$

and $\bar{n} = \bar{n}_0 + 1$.

Denote the numerator and denominator of $\hat{C}_i(q^{-1})$ and $\hat{G}_i(q^{-1})$ as $(\hat{N}_{c,i}(q^{-1}), \hat{D}_{c,i}(q^{-1}))$ and $(\hat{N}_{p,i}(q^{-1}), \hat{D}_{p,i}(q^{-1}))$, respectively, i.e.,

$$\hat{C}_i(q^{-1}) = \frac{\hat{N}_{c,i}(q^{-1})}{\hat{D}_{c,i}(q^{-1})}; \quad \hat{G}_i(q^{-1}) = \frac{\hat{N}_{p,i}(q^{-1})}{\hat{D}_{p,i}(q^{-1})} \qquad (9.5.16)$$

The pole-assignment algorithm then calculates $(\hat{N}_{c,i}(q^{-1}), \hat{D}_{c,i}(q^{-1}))$ such that

$$\hat{N}_{c,i}(q^{-1})\hat{N}_{p,i}(q^{-1}) + \hat{D}_{c,i}(q^{-1})\hat{D}_{p,i}(q^{-1}) = T(q^{-1}) \qquad (9.5.17)$$

holds. For example, if we further denote

$$\hat{N}_{c,i}(q^{-1}) = \sum_{i=1}^{n_c-1} c_i q^{-i} \qquad (9.5.18)$$

$$\hat{D}_{c,i}(q^{-1}) = \sum_{i=1}^{n_c} d_i q^{-i} \qquad (9.5.19)$$

$$\hat{N}_{p,i}(q^{-1}) = \sum_{i=1}^{n_p-1} g_i q^{-i} \qquad (9.5.20)$$

$$\hat{D}_{p,i}(q^{-1}) = \sum_{i=1}^{n_p} f_i q^{-i} \qquad (9.5.21)$$

where n_c and n_p are integers satisfying $n_c + n_p = \bar{n}$ then by expressing

$$T(q^{-1}) = 1 + h_1 q^{-1} + \cdots + h_{\bar{n}} q^{-\bar{n}} \qquad (9.5.22)$$

the following equation can be obtained from Eq. (9.5.17).

$$\Omega \theta = H \qquad (9.5.23)$$

where

$$\Omega = \begin{bmatrix} g_0 & 0 & 0 & \cdot & \cdot & \cdot & 0 & 1 & 0 & 0 & \cdot & \cdot & \cdot & 0 \\ g_1 & g_0 & 0 & \cdot & \cdot & \cdot & 0 & f_1 & 1 & 0 & \cdot & \cdot & \cdot & 0 \\ \cdot & \cdot & \cdot & \cdot & \cdot & \cdot & \cdot & \cdot & \cdot & \cdot & \cdot & \cdot & \cdot & \cdot \\ \cdot & \cdot & \cdot & \cdot & \cdot & \cdot & \cdot & \cdot & \cdot & \cdot & \cdot & \cdot & \cdot & \cdot \\ \cdot & \cdot & \cdot & \cdot & \cdot & \cdot & \cdot & \cdot & \cdot & \cdot & \cdot & \cdot & \cdot & \cdot \\ g_{n_p} & g_{n_p-1} & \cdot & \cdot & \cdot & \cdot & g_0 & f_{n_p} & f_{n_p-1} & \cdot & \cdot & \cdot & f_1 & 1 \\ 0 & g_{n_p} & g_{n_p-1} & \cdot & \cdot & \cdot & & g_0 & f_{n_p} & f_{n_p-1} & \cdot & \cdot & \cdot & f_1 \\ \cdot & \cdot & \cdot & \cdot & \cdot & \cdot & \cdot & \cdot & \cdot & \cdot & \cdot & \cdot & \cdot & \cdot \\ \cdot & \cdot & \cdot & \cdot & \cdot & \cdot & \cdot & \cdot & \cdot & \cdot & \cdot & \cdot & \cdot & \cdot \\ 0 & 0 & 0 & \cdot & \cdot & \cdot & g_{n_p} & 0 & 0 & 0 & \cdot & \cdot & \cdot & f_{n_p} \end{bmatrix}$$

(9.5.24)

$$\theta^T = [\, c_0, \quad c_1, \quad \cdots, \quad c_{n_c}, \quad 1, \quad d_1, \quad \cdots, \quad d_{n_c}\,]$$

(9.5.25)

$$H^T = [\, 1, \quad h_1, \quad \cdots, \quad h_{\tilde{n}}\,]$$

(9.5.26)

Therefore the parameters of controller c_i and d_i can be directly obtained by solving Eq. (9.5.23).

vi) Repeat ii)-v) until

$$\left| \hat{G}_i(q^{-1}) - \hat{G}_{k-1}(q^{-1}) \right| \leq \delta$$

(9.5.27)

where $\delta \ll 1$.

This approach is similar to that of indirect self-tuning control. The difference here is that the rule based identification algorithm is used to identify the plant and to tune the gains of the controller. The rule based identification described in sections 9.2-9.4 is based on a step response test of the open loop system, however the rule based tuner is an on-line method which requires the on-line identification of the closed-loop system. To allow the rule based identification to be used on-line it is applied between the input signal and the integral of the tracking error.

Since the rule based identification method of Wang and Jones [149] has shown to be convergent by only using single step input, the above tuning procedure can achieve pre-specified dynamic performance within a single cycle in Fig. 9.9. However, if the structure of the controller is changed during the tuning phase, a further step input has to be applied and consequently, several plant step inputs are needed. Therefore, for some plants (especially long time constant plant), an accurate choice of the initial structure for the controller should be made in order to achieve fast tuning.

It should also be noted that the convergence of the integral of error is used in iv) above to determine the number of integrators required rather than the identified model in order to avoid problems caused by slight estimation errors. Also, in order to obtain a solution to the diophantine Eq. (9.5.17) it may be necessary to update the controller in Eq. (9.5.10) without including the integrator term.

Compared with the widely used closed loop system identification approaches in which the input and output of the plant are directly used ([104, 57]), the closed loop system identification in stage iii) of the above algorithm only uses the demand input and the tracking error. As a result there is no requirement on the closed-loop identifiability.

9.6 Simulation and Application

9.6.1 Rule based modelling

In this subsection, we will consider the application of the rule based identification to the modelling of a velocity control system for a jet engine. The input to the system is the engine fuel flow rate and the output from the system is the spool speed of the engine. The system is single-input and single-output. The velocity data are generated by performing a step response test on the jet engine test model, and are stored in the data base of the program. From real data, which are corrupted by noise, it can be seen that the velocity control system is of the MPM type. Figure 9.10 shows the results of identification, giving the system model as follows:

$$M(\hat{d}, \hat{n}, \hat{m}, \hat{p}) = \frac{q^{-1}(0.199 - 0.067q^{-1})}{1 - 1.72q^{-1} + 0.739q^{-2}} \tag{9.6.1}$$

with time delay equal to one sampling period and the structure orders $\hat{n} = 2$ and $\hat{m} = 1$. The corresponding performance function J is equal to 0.0268. From Fig. 9.10, it can be seen that the matching model gives a good estimation for the original system.

Our second example is concerned with a coupled electric drives shown in Fig. 9.11. The system is multivariable representing a number of industrial systems. The rig consists of a continuous elastic belt which is fitted around two pulleys driven by identical servomotors. A third jockey pulley is spring loaded. The system inputs are voltage signals to the servomotors and the outputs are the belt tension and the angular velocity of the jockey pulley. The discrete time model is therefore given by

$$\omega(k) = S_{11}(q^{-1})u_1(k) + S_{12}(q^{-1})u_2(k) \tag{9.6.2}$$
$$T_n(k) = S_{21}(q^{-1})u_1(k) + S_{22}(q^{-1})u_2(k) \tag{9.6.3}$$

where, in practice, $\omega(k)$, $T_n(k)$ and $u_i(k)$ are the angular velocity, the belt tension and the voltage signal to the ith drive, respectively. The problem is to use the rule based identification algorithm to estimate the unknown system transfer functions, $S_{11}(q^{-1}); i, j = 1, 2$.

Again, the data are generated by applying an input signal to one drive at a time and noting the output responses. Results are only presented for the transfer matrix elements

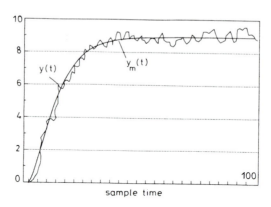

Figure 9.10: Modelling results.

corresponding to drive 1 as the results for drive 2 are similar. Application of the rule
based identification yields the matching models

$$M_{11}(q^{-1}) = \frac{q^{-1}(0.098 - 0.03q^{-1})}{1 - 1.60q^{-1} + 0.64q^{-2}} \qquad (9.6.4)$$

$$M_{21}(q^{-1}) = \frac{q^{-1}(0.143 + 0.048q^{-1})}{1 - 2.74q^{-1} + 3.29q^{-2} - 1.95q^{-3} + 0.51q^{-4}} \qquad (9.6.5)$$

Figures 9.12-9.13 shows the identification results; where Figure 9.12 corresponds to the
velocity model (S_{11}, M_{11}) whilst Figure 9.13 corresponds to the tension model (S_{21}, M_{21}).
Again, desired results are obtained. Detailed comparison of these results with the instrumental variable (IV) method was carried out by Daley and Wang in 1991[38] and
it has been shown that the rule based approach is a valuable tool for providing a quick
estimation on model structures and parameters.

9.6.2   Rule based control

The simulation study considered here is the process trainer PT326 manufactured by Feedback which is widely used in universities for teaching purposes. As shown in Fig 9.14, air
is driven, by a centrifugal blower, past a heater grid and through a length of tubing. The
aim is to control the temperature of the air flowing through the tube, and it is desired
that the controller brings the temperature to the set point value as quickly as possible.

When the temperature sensor is close to the left end of the tube and the sampling
time is set to 0.1 second, the transfer function for the process can be expressed by

$$\hat{G}_i(q^{-1}) = \frac{0.0539}{1 - 0.951q^{-1}} \qquad (9.6.6)$$

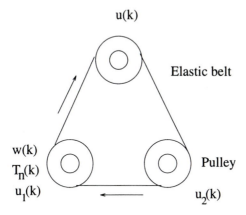

Figure 9.11: The structure of the coupled electric drives.

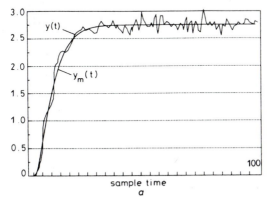

Figure 9.12: Modelling results for the velocity system (S_{11}, M_{11}).

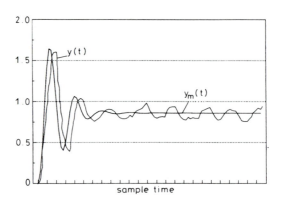

Figure 9.13: Modelling results for the tension system (S_{21}, M_{21}).

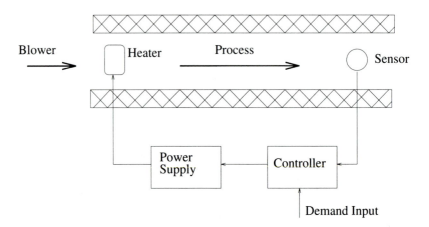

Figure 9.14: PT326 process trainer.

where the proportional band is set to 100%. In this case, the initial structure of the controller is proportional, i.e.,

$$\hat{C}_0(q^{-1}) = 0.5 \tag{9.6.7}$$

and the period of the demand input is $T_0 = 4$ seconds. The desired polynomial $T(q^{-1})$ is chosen as follows

$$T(q^{-1}) = 1 - 1.083q^{-1} + 0.14229q^{-2} = (1 - 0.93q^{-1})(1 - 0.153q^{-1}) \tag{9.6.8}$$

where the pole $\lambda_1 = 0.93$ is dominant and $\lambda_2 = 0.153$ is a non-dominant. Since both poles are real, the overshoot corresponding to the desired closed loop response is 0% with the setting time (less than 2.0 seconds) being determined by the dominant pole. These form the requirements for dynamic performance. In the simulation, the sampling period is set to 0.1 seconds. With the application of the controller tuning algorithm developed in the previous sections, the final structure of the control is a PI controller and

$$\hat{C}_5(q^{-1}) = 8.72 + \frac{0.098}{1 - q^{-1}} \tag{9.6.9}$$

The tuning phase is shown in Fig .9.15 in which the transient response of the system is given, and after $i = 3$ a good response satisfying the requirements of dynamic performance (0% overshoot and less than 2 second setting time) are achieved. In Fig 9.16, the controller gains during the tuning phase is shown.

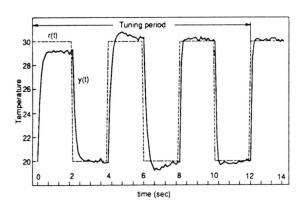

Figure 9.15: The closed loop system response.

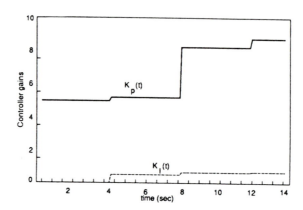

Figure 9.16: The tuning phase of the control parameters.

9.7 Conclusions

In this chapter, we have discussed rule-based identification and controller tuning. It can be seen that the rule based identification is mainly off-line. However, compared with the most developed recursive identification methods used in self-tuning adaptive control systems, the following advantages are obvious for the rule based method:

- Desired estimation results can be obtained without any restriction on the noise type;

- The time delay and the structure orders of the systems can be estimated simultaneously;

- There is good robustness with respect to abrupt noise changes;

- Only one open-loop step response test on the real system is needed.

Using the rule based identification algorithm, a new type of controller tuning algorithm has been described. With direct identification from the measurements of the demand input and the integrated tracking error, the rule based identification algorithm can be used on-line. During the tuning phase, both the gains and the structure of the controller are updated in order to stabilize the closed loop system, achieve desired dynamic performance and realize perfect tracking.

Application to various processes demonstrates the potential ability of the rule based adaptive tuning method. Moreover, it can be seen that the controller tuning algorithm described in this chapter can also be applied to time-varying systems if the tuner is switched on when the change inside the process is detected. Analytical redundancy based approaches are available to detect such changes (Willsky, [170] and Isermann, [69]), however, with the rule based identification method and under the assumption that a fine tuned controller has already been obtained, the detection can be simply carried out in the following ways:

- Apply a step input to the closed loop system in regular intervals to monitoring the dynamic performance, such as the overshoot and the setting time;

- If there is an apparent change, switch on the rule based tuner and apply a number of step inputs (Fig. 9.9) to the closed loop system to retune the controller.

Difficulty may arise when applying step inputs regularly to practical systems, since some of them cannot be accommodated with testing signals. This is a common limitation of the testing signal based approaches and in practise, one should keep the magnitude of the testing signal to a minimum.

Since it has been shown by Wang and Jones [149] that the rule based identification algorithm can be directly used to identify the multi-input and multi-output (MIMO) systems, the control tuning algorithm described in this chapter can also be used for MIMO system control. In this case, the individual component in the demand input $y_r(k)$ vector should satisfy non-overlap conditions, this means that at any time either $y_r(k) = 0$ or there is only one component whose value is 1.

Chapter 10

Adaptive Control of Singular Systems

10.1 Introduction

In practice, many system models can be expressed by the composition of a set of dynamic and static relations, resulting in a need for representation for singular systems. An example of such a system is the electrical network ([147]) shown in Fig. 10.1, where the two capacitors are connected in parallel and are subject to the input current $i(t)$.

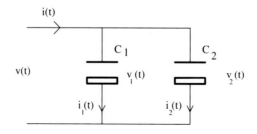

Figure 10.1: An electrical network.

The output of the electrical network is the voltage $v(t)$ across the connected capacitors. Assuming that the values of the capacitors are $C_1 = C_2 = 1$ and using the notations given in Fig. 10.1, the following equations can be directly obtained.

$$\dot{v}_1(t) = i_1(t) \tag{10.1.1}$$
$$\dot{v}_2(t) = i_2(t) \tag{10.1.2}$$
$$v_1(t) = v_2(t) \tag{10.1.3}$$
$$i_1(t) + i_2(t) = i(t) \tag{10.1.4}$$

$$v(t) \quad = \quad v_2(t) \tag{10.1.5}$$

which can be further expressed in matrix form as

$$E\dot{x}(t) \quad = \quad Ax(t) + Bu(t) \tag{10.1.6}$$
$$y(t) \quad = \quad Cx(t) \tag{10.1.7}$$

where

$$E = \begin{bmatrix} 1 & 0 & 0 & 0 \\ 0 & 1 & 0 & 0 \\ 0 & 0 & 0 & 0 \\ 0 & 0 & 0 & 0 \end{bmatrix} ; A = \begin{bmatrix} 0 & 0 & 1 & 0 \\ 0 & 0 & 0 & 1 \\ 1 & -1 & 0 & 0 \\ 0 & 0 & 1 & 1 \end{bmatrix}$$

$$B = \begin{bmatrix} 0 \\ 0 \\ 0 \\ -1 \end{bmatrix} ; C = \begin{bmatrix} 1 & 0 & 0 & 0 \end{bmatrix}$$

Fundamental research on the implementation and design of linear singular control systems has been carried out for many years ([147, 174, 36, 37]). However, the majority approaches were addressed to known singular systems and the first adaptive control scheme for singular systems was developed only recently by Wang and Daley in 1993 ([151]). The application of their method ([40]) to a gas turbine control has been shown to be very successful. Adaptive control for singular systems belongs to the scope of advanced adaptive control and is therefore described in details in this chapter. Our presentation will only concentrate on the adaptive control of linear discrete time singular systems.

10.2   System Representation

Consider the single-input single-output singular state space model of a singular system

$$Ex(k+1) \quad = \quad Ax(k) + Bu(k) \tag{10.2.1}$$
$$y(k) \quad = \quad Cx(k) \tag{10.2.2}$$

where $x(k) \in R^n$ is the state vector, $u(k) \in R^1$ is the input signal and $y(k) \in R^1$ is the output signal. E, A, B and C are unknown parameter matrices of appropriate dimensions and $det(E) = 0$. For conventional systems E is equal to the unit matrix, for singular systems, (10.2.1) and (10.2.2) can be either causal or non-causal. This specific characteristic has been widely studied in the literature ([147, 174, 36, 37]).

Definition 10.1 The singular system (10.2.1) and (10.2.2) is regular if

$$det(\lambda E - A) \neq 0 \tag{10.2.3}$$

Eq.(10.2.3) is therefore referred to as the condition of regularity. It has been shown that,under the condition of regularity, two nonsingular matrices $P \in R^{n \times n}$ and $Q \in R^{n \times n}$ can be constructed ([147]) such that

$$QEP = \begin{bmatrix} I_{n_1} & 0 \\ 0 & N \end{bmatrix} \tag{10.2.4}$$

$$QAP = \begin{bmatrix} A_1 & 0 \\ 0 & I_{n_2} \end{bmatrix} \tag{10.2.5}$$

$$QB = \begin{bmatrix} B_1 \\ B_2 \end{bmatrix} \tag{10.2.6}$$

$$x(k) = P \begin{bmatrix} x_1(k) \\ x_2(k) \end{bmatrix} \tag{10.2.7}$$

where N is a nilpotent matrix of order l, i.e.,

$$N^{l-1} \neq 0, N^l = 0 \tag{10.2.8}$$

As a result, the singular system (10.2.1) and (10.2.2) can be equivalently expressed by the following form

$$x_1(k+1) = A_1 x_1(k) + B_1 u(k) \tag{10.2.9}$$
$$N x_2(k+1) = x_2(k) + B_2 u(k) \tag{10.2.10}$$
$$y(k) = C_1 x_1(k) + C_2 x_2(k) \tag{10.2.11}$$

In the literature, system (10.2.9)-(10.2.11) is referred to as a Restricted System Equivalent (RSE) to system (10.2.1)-(10.2.2).

Example 10.1 Consider the following singular system

$$Ex(k+1) = Ax(k) + Bu(k) \tag{10.2.12}$$
$$y(k) = Cx(k) \tag{10.2.13}$$

with

$$E = \begin{bmatrix} 1 & 1 \\ 0 & 0 \end{bmatrix}; A = \begin{bmatrix} 0 & 0 \\ 0 & 1 \end{bmatrix}$$

$$B = \begin{bmatrix} 0 \\ 1 \end{bmatrix}; C = \begin{bmatrix} 1 & 0 \end{bmatrix}$$

It can be seen that the following two matrices

$$Q = \begin{bmatrix} 1 & -1 \\ 0 & 1 \end{bmatrix}; P = \begin{bmatrix} 1 & 0 \\ 0 & 1 \end{bmatrix}$$

will transform the original system into

$$x_1(k+1) = -u(k) \tag{10.2.14}$$
$$0 = x_2(k) + u(k) \tag{10.2.15}$$
$$y(k) = x_1(k) \tag{10.2.16}$$

For the system (10.2.9)-(10.2.11), the solution for $x_2(k)$ can be expressed as

$$x_2(k) = \sum_{j=0}^{l-1} N^j B_2 u(k+j) \tag{10.2.17}$$

By substituting Eq. (10.2.17) into Eq. (10.2.11), the system output $y(k)$ can be further expressed as

$$y(k) = C_1 x_1(k) + \sum_{j=0}^{l-1} C_2 N^j B_2 u(k+j) \tag{10.2.18}$$

From Eq. (10.2.18) it can be seen that the output of the system (10.2.9) -(10.2.11) (or (10.2.1) and (10.2.2)) at time k not only depends on the input

$$\{u(k), u(k-1), \ldots, u(0)\} \tag{10.2.19}$$

but also on the future input

$$\{u(k+1), u(k+2), \ldots, u(k+l-1)\} \tag{10.2.20}$$

Since the future input is not available at time k, the system is non-causal and it is difficult to design an adaptive controller using standard techniques. It can actually be seen that the system will be causal if and only if

$$C_2 B_2 = C_2 N B_2 = \cdots = C_2 N^{l-1} B_2 = 0 \tag{10.2.21}$$

Moreover, as we have already discussed before, general linear identification algorithms for unknown parameters are based upon the assumption that the system can be expressed in the form of

$$y(k) = \theta^T \phi(k-1) \tag{10.2.22}$$

where both the output $y(k)$ and the measurable vector $\phi(k-1)$ are required to be available at time k. However, this is clearly not true for non-causal systems, as future values for $u(k)$ can not be included in $\phi(k-1)$.

10.3   Parameter Estimation

Instead of identifying the unknown parameter matrices A, B, C and E directly, the unknown parameters of the input-output form are estimated. Therefore, it is necessary to formulate the input-output expression of the system (10.2.1)-(10.2.2). From Eqs. (10.2.9)-(10.2.11), it can be seen that the output of the original system (10.2.1)-(10.2.2) can be expressed as

$$\begin{aligned} y(k) &= C_1 (I_{n_1} - q^{-1} A_1)^{-1} B_1 u(k) + \\ &\quad + \sum_{j=0}^{l-1} C_2 N^j B_2 u(k+j) \end{aligned} \tag{10.3.1}$$

Therefore, the following input-output expression can be obtained.

$$\eta_m y(k+m) + \eta_{m-1} y(k+m-1) + \cdots + \eta_0 y(k) +$$
$$+\beta_m u(k+m) + \beta_{m-1} u(k+m-1) + \cdots + \beta_0 u(k) = 0 \qquad (10.3.2)$$

where m is an integer which is less than or equal to the system order n and $(\eta_0, \eta_1, \cdots, \eta_m, \beta_0, \beta_1, \cdots, \beta_m)$ are unknown coefficients which depend on the actual values of A, B, C and E. This is a general form and can be used to represent both causal and non-causal systems. For example, if $\eta_m \neq 0$, then the original system is causal, otherwise, the original system is non-causal. Since an integer m always exists such that $\beta_m \neq 0$, without loss of generality, it can be assumed that $\beta_m = 1$ and the system (10.3.1) can be expressed in the following alternative form

$$u(k+m) = \theta^T \phi(k+m) \qquad (10.3.3)$$

where

$$\theta^T = [-\eta_m, -\eta_{m-1}, \cdots, -\eta_0,$$
$$-\beta_{m-1}, -\beta_{m-2}, \cdots, -\beta_0] \in R^{2m+1} \qquad (10.3.4)$$
$$\phi(k+m) = [y(k+m), y(k+m-1), \cdots, y(k), u(k+m-1),$$
$$u(k+m-2), \cdots, u(k+1), u(k)] \in R^{2m+1} \qquad (10.3.5)$$

By taking m backward steps on both sides of Eq. (10.3.3), we have

$$u(k) = \theta^T \phi(k) \qquad (10.3.6)$$

Since at time k the vector $\phi(k)$ is available, this equation can be used to construct a recursive identification algorithm. Denote $\hat{\theta}(k)$ as the estimated value of θ at time k

$$\hat{\theta}(k) = [-\hat{\eta}_m(k), -\hat{\eta}_{m-1}(k), \cdots, -\hat{\eta}_0(k), -\hat{\beta}_{m-1}(k), \cdots, -\hat{\beta}_0(k)] \in R^{2m+1} \qquad (10.3.7)$$

where $\hat{\eta}_i(k)$ and $\hat{\beta}_j(k)$ are the estimates of η_i and β_j, respectively. Similar to ordinary recursive least square algorithm, let us define the residual signal as

$$\epsilon(k) = u(k) - \hat{\theta}^T(k-1)\phi(k) \qquad (10.3.8)$$

then the recursive least squares identification algorithm for unknown singular system can be constructed as follows

$$\Delta\hat{\theta}(k) = \hat{\theta}(k) - \hat{\theta}(k-1) = \frac{P(k-2)\phi(k)\epsilon(k)}{1 + \phi^T(k)P(k-2)\phi(k)} \qquad (10.3.9)$$
$$\epsilon(k) = u(k) - \hat{u}(k) \qquad (10.3.10)$$
$$\hat{u}(k) = \hat{\theta}^T(k-1)\phi(k) \qquad (10.3.11)$$
$$P^{-1}(k-1) = P^{-1}(k-2) + \phi(k)\phi^T(k) \qquad (10.3.12)$$

where $\hat{\theta}(0) = 0$ and $P(0) = P_0 > 0$ are the initial values of $\hat{\theta}(k)$ and $P(k)$.

Compared with the ordinary least squares identification algorithm in chapter 2, it can be seen that here, the residual signal has a different form and it can be used to identify both causal and non-causal system. For example, if the system is causal and has pure time delay $d > 0$, the residual signal can take the form of

$$\epsilon(k) = u(k - d) - \hat{\theta}^T(k - 1)\phi(k) \qquad (10.3.13)$$

with

$$\phi(k) = [y(k), y(k - 1), \cdots, y(k - m + 1), y(k - m),$$
$$u(k - d - 1), u(k - d - 2), \cdots, u(k - d - m)] \in R^{2m+1} \qquad (10.3.14)$$

Moreover, the algorithm (10.3.9)-(10.3.12) can be used to on-line check the causality of the unknown singular system. For instance, it can be concluded that the original system is non-causal if the first estimated element $\hat{\eta}_m$ in $\hat{\theta}(k)$ converges to zero, as demonstrated by the following simulation examples.

Example 10.2 Consider the singular system

$$Ex(k + 1) = Ax(k) + Bu(k) \qquad (10.3.15)$$
$$y(k) = Cx(k) \qquad (10.3.16)$$

where

$$A = \begin{bmatrix} 0 & 1 & 0 & 0 & 0 \\ 0 & 0 & 1 & 0 & 0 \\ 0 & 0 & 1 & 0 & 0 \\ 0 & 0 & 0 & 1 & 0 \\ 0 & 0 & 0 & 0 & 1 \end{bmatrix} \qquad (10.3.17)$$

$$E = \begin{bmatrix} 1 & 0 & 0 & 0 & 0 \\ 0 & 1 & 0 & 0 & 0 \\ 0 & 0 & 1 & 0 & 0 \\ 0 & 0 & 0 & 0 & 1 \\ 0 & 0 & 0 & 0 & 0 \end{bmatrix} \qquad (10.3.18)$$

$$B = \begin{bmatrix} 0 & 0 & 1 & 1 & 1 \end{bmatrix}^T \qquad (10.3.19)$$
$$C = \begin{bmatrix} 1 & 0 & 0 & 0 & 1 \end{bmatrix} \qquad (10.3.20)$$

It can be directly formulated that the input-output relation of this system is

$$y(k) - y(k - 1) - u(k) + u(k + 1) - u(k - 2) = 0 \qquad (10.3.21)$$

which is, of course, causal. Let $m = 4$, then the vector $\phi(k)$ is of the form

$$\phi(k) = [y(k), y(k - 1), y(k - 2), y(k - 3), y(k - 4)$$
$$u(k - 1), u(k - 2), u(k - 3), u(k - 4)]^T \in R^9 \qquad (10.3.22)$$

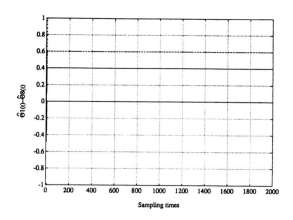

Figure 10.2: Parameter estimation for a causal system.

The system is subject to a pseudo square wave with amplitude $\{\pm 1\}$, the identification results are shown in Fig. 10.2.

It can be seen that the estimates rapidly converge to their true values and the corresponding $\hat{\theta}(k)$ when $k = 1000$ becomes

$$\hat{\theta}(1000) = [-1 \quad 0.6 \quad 0.4 \quad 0 \quad 0 \quad 0.6 \quad 0.4 \quad 1 \quad 0.4]^{T} \qquad (10.3.23)$$

Therefore the system can be approximately expressed as

$$u(k) = \hat{\theta}^{T}\phi(k) = -y(k) + 0.6y(k-1) + 0.4y(k-2) + 0.6u(k-1)$$
$$+0.4u(k-2) + u(k-3) + 0.4u(k-4) \qquad (10.3.24)$$

By introducing the backward shift operator, q^{-1}, of chapter 2, it can be obtained that

$$(1 - 0.6q^{-1} - 0.4q^{-2})y(k) \quad = \quad (-1 + 0.6q^{-1}$$
$$+0.4q^{-2} + q^{-3} + 0.4q^{-4})u(k) \qquad (10.3.25)$$

It can be observed that the two polynomials in q^{-1} have a common factor $(1 + 0.4q^{-1})$ which has a stable zero, -0.4. By cancelling this common factor, the input-output form

$$y(k) - y(k-1) - u(k) + u(k-1) - u(k-2) = 0 \qquad (10.3.26)$$

can be obtained which coincides with the original algebraically expressed form of Eq. (10.3.21).

Example 10.3 Re-consider the previous example with the same matrices A, B and E but with a C defined by

$$C = [1 \quad 0 \quad 0 \quad 1 \quad 0] \qquad (10.3.27)$$

Similar to Example 10.2, it can be shown that this is a non-causal system. The input-output relationship in this case is given by

$$y(k-1) - y(k-2) + u(k) + u(k-1) - 2u(k-2) - u(k-4) = 0 \qquad (10.3.28)$$

The identification results for this system are shown in Fig. 10.3 with

$$\begin{aligned}\hat{\theta}(1000) &= [-0.0001, -0.9979, 0.9979, 0.001, 0 - 0.9999 \\ &\quad 1.9969, 0.0031, 0.9990]^T \end{aligned} \qquad (10.3.29)$$

which corresponds to the estimated model

$$0.0001y(k) + 0.9979y(k-1) - 0.9979y(k-2) - 0.0010y(k-3) + u(k)$$
$$+0.9999u(k-1) - 1.9969u(k-2) - 0.0031u(k-3) - 0.9990u(k-4) = 0 \quad (10.3.30)$$

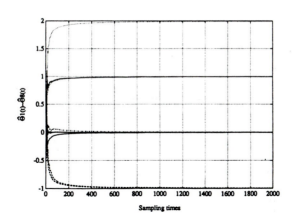

Figure 10.3: Parameter estimation for a non-causal system.

Obviously, the accuracy of the parameter estimates affects the conclusion regarding the causality of the unknown system, therefore the usual requirement that the input signal be persistently exciting is necessary in order to ensure good parameter convergence. If necessary, the exponentially decreased persistently exciting signal proposed by Chen and Guo in 1986 ([23]) should be utilized for this purpose. In the above examples, the input signal was persistently exciting and hence desired convergence of the estimates is obtained.

The accuracy of the estimates will also be affected by the presence of measurement noise. An appropriate identification algorithm (Extended Least Squares (ELS), Instrumental Variables (IV) for example) must be used if the system is subjected to noise[41].

10.4 Preliminary Feedback Design

As the estimated singular system may be either causal or non-causal, conventional adaptive control is not applicable. The idea is therefore to construct a preliminary output feedback to make the system causal. In this section, the realization of the state space form and the design of a preliminary output feedback gain will be described. To simplify the formulation, the following definition ([37]) is used.

Definition 10.2 The system (10.2.1)-(10.2.2) is controllable and observable if and only if

$$rank[\lambda E - A, B] = rank[\lambda E - A/C] = n \qquad (10.4.1)$$

Using this definition, the following lemma can be obtained. Its proof can be found in [37].

Lemma 10.1 *Assuming that the singular system (10.2.1)-(10.2.2) is controllable and observable, then there will always exists an output feedback*

$$u(k) = Gy(k) + v(k) \qquad (10.4.2)$$

such that the input-output causality relationship holds between $v(k)$ and $y(k)$.

Therefore, if this feedback relation can be found with the help of the parameter estimation results, the adaptive control algorithm can be designed based on the relationship between $v(k)$ and $y(k)$. The feedback (10.4.2) is called preliminary output feedback whose only purpose is to achieve a causal relationship between $v(k)$ and $y(k)$. Since for causal systems the feedback gain matrix, G, is zero, then in the following only non-causal systems will be considered.

Based upon the estimation result, the original system can be expressed by

$$
\begin{aligned}
u(k+m) &= \hat{\theta}(k+m)\phi(k+m) + \epsilon(k+m) \\
&= \hat{\theta}^T(k)\phi(k+m) + [\hat{\theta}(k+m) - \hat{\theta}(k)]^T\phi(k+m) + \epsilon(k+m) \quad (10.4.3)
\end{aligned}
$$

Denote

$$\sigma(k+m) = \hat{\theta}(k+m) - \hat{\theta}(k)]^T\phi(k+m) \qquad (10.4.4)$$

Then we have

$$u(k+m) = \hat{\theta}^T(k+m)\phi(k+m) + \sigma(k+m) + \epsilon(k+m) \qquad (10.4.5)$$

Assume that with the identification results, there exists an $m_0 < m$ such that the estimates for $\eta_m, \eta_{m-1}, \cdots, \eta_{m_0+1}$ converge to zero, i.e. the original system is non-causal. Then the original system can be expressed by

$$
\begin{aligned}
\hat{\eta}_{m_0}(k)y(t+m_0) + \hat{\eta}_{m_0-1}(k)y(k+m_0-1) + \cdots + \hat{\eta}_1(k)y(k-1) + \hat{\eta}_0 y(k) = \\
+u(k+m) + \hat{\beta}_{m-1}(k)u(k+m-1) + \cdots + \hat{\beta}_1(k)u(k+1) + \hat{\beta}_0(k)u(k) = \\
= \sigma(k+m) + \epsilon(k+m) \quad (10.4.6)
\end{aligned}
$$

which yields the following state space realization

$$E_r z(k+1) = A_r(k)z(k) + B_r u(k) \tag{10.4.7}$$
$$y(k) = C_r z(k) + \gamma(k) \tag{10.4.8}$$

where

$$E_r = \begin{bmatrix} I_{m_0} & 0 & 0 \\ 0 & I_{m_0} & 0 \\ 0 & 0 & N_{k-m_0+1} \end{bmatrix} ; A_r = \begin{bmatrix} \hat{A}_1 & 0 & 0 \\ 0 & \hat{A}_2 & \hat{B}_{21}\hat{C}_2 \\ 0 & 0 & I_{k-m_0+1} \end{bmatrix} \tag{10.4.9}$$

$$B_r = \begin{bmatrix} \hat{B}_1 \\ 0 \\ \hat{B}_2 \end{bmatrix} ; C_r = [\hat{C}_1 \quad \hat{C}_{21} \quad 0] ; N_{k-m_0+1} = \begin{bmatrix} 0 & I_{k-m_0} \\ 0 & 0 \end{bmatrix} \tag{10.4.10}$$

$$\hat{A}_1 = \hat{A}_2 = \begin{bmatrix} 0 & I_{m_0-1} \\ -\dfrac{\hat{\eta}_0(k)}{\hat{\eta}_{m_0-1}(k)} & \xi \end{bmatrix} \tag{10.4.11}$$

$$\hat{B}_{21} = [0,0,\cdots,0-\frac{1}{\hat{\eta}_{m_0}(k)}]^T \tag{10.4.12}$$

$$\hat{B}_1 = -\frac{1}{\hat{\eta}_{m_0}}[\hat{\beta}_0(k),\hat{\beta}_1(k),\cdots,\hat{\beta}_{m_0-1}]^T \tag{10.4.13}$$

$$\hat{C}_1 = \hat{C}_{21} = [1,0,0,\cdots,0] \in R^{1\times m_0} \tag{10.4.14}$$

$$\hat{C}_2 = [1,0,0,\cdots,0,0] \in R^{m-m_0+1} \tag{10.4.15}$$

$$\xi = -\frac{1}{\hat{\eta}_{m_0}(k)}[\hat{\eta}_1(k),\hat{\eta}_2(k),\cdots,\hat{\eta}_{m_0-1}(k)] \tag{10.4.16}$$

$$\gamma(k) = \frac{\sigma(k+m-m_0)+\epsilon(k+m-m_0)}{\hat{\eta}_{m_0}(k)+\hat{\eta}_{m_0-1}(k)q^{-1}+\cdots+\hat{\eta}_0(k)q^{-m_0}} \tag{10.4.17}$$

By making use of the properties of the least squares algorithm (see chapter 2), it can be seen that

$$\gamma(k) = o(\phi(k+m-m_0)) \tag{10.4.18}$$

that is

$$\lim_{k\to\infty} \frac{\gamma(k)}{\phi(k+m-m_0)} = 0 \tag{10.4.19}$$

and therefore, $\gamma(k)$ has little influence on the dynamic behaviour when k is large.

With the aid of Definition 10.1, it is easy to see that the system (10.4.7) and (10.4.8) is both controllable and observable. Therefore, by utilizing Lemma 10.1, it can be concluded that there always exists an output feedback gain G such that the preliminary closed loop system

$$E_r z(k+1) = (A_r + B_r GC_r)z(k) + B_r v(k) \tag{10.4.20}$$
$$y(k) = C_r z(k) + \gamma(k) \tag{10.4.21}$$

is input-output causal. The calculation of the gain G is carried out as follows (Dai, 1989; [37]):

i) Let P_r and Q_r be non-singular matrices such that

$$Q_r E_r P_r = \begin{bmatrix} I_h & 0 \\ 0 & 0 \end{bmatrix} ; Q_r A_r P_r = \begin{bmatrix} A_{11} & A_{12} \\ A_{21} & A_{22} \end{bmatrix} \tag{10.4.22}$$

$$Q_r B_r = \begin{bmatrix} B_1 \\ B_2 \end{bmatrix} ; C_r P_r = [C_1 \quad C_2] \tag{10.4.23}$$

ii) Choose

$$G = B_2^T (B_2 B_2^T)^{-1} (I - A_{22}) (C_2^T C_2)^{-1} C_2^T \tag{10.4.24}$$

It should be noted that the above choice of gain matrix G is only a sufficient condition for the preliminary output feedback system (10.4.20)-(10.4.21) to be causal. In some cases, a fixed gain matrix G can also achieve this aim. Therefore, the choice of the preliminary feedback gain is not limited to the above form.

10.5   Adaptive Control Design

Since the relationship between $v(k)$ and $y(k)$ is causal, any classical adaptive control algorithm can be directly designed for the system , such as the d-step-ahead-predictive control ([57]) or the pole assignment algorithm ([163]). The design procedure for adaptive controller is therefore summarized as follows:

1) Use the algorithm (10.3.9)-(10.3.12) to identify the unknown parameters and decide whether the original system is causal or not;

2) If the original system is causal, set the preliminary gain $G = 0$, otherwise, calculate the gain G as given in (10.4.24);

3) Design a controller for the preliminary closed loop system (10.4.20)-(10.4.21).

This process can be repeated continuously for time-varying systems. The general structure of the closed loop system is shown in Fig. 10.4, where $y_r(k)$ is the reference input. From this structure it is clear that all the current adaptive control algorithms can be combined with step 1) and 2).

10.6   Simulation Results

Consider the singular system

$$Ex(k+1) = Ax(k) + Bu(k) \tag{10.6.1}$$
$$y(k) = Cx(k) \tag{10.6.2}$$

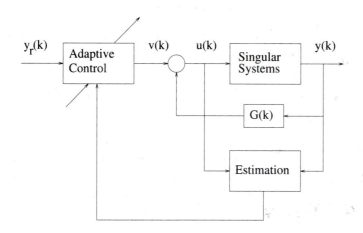

Figure 10.4: The structure of the closed-loop system.

with

$$A = \begin{bmatrix} 0.5 & 0 & 0 \\ 0 & 1 & 0 \\ 0 & 0 & 1 \end{bmatrix} ; E = \begin{bmatrix} 1 & 1 & 0 \\ 0 & 0 & 1 \\ 0 & 0 & 0 \end{bmatrix} \tag{10.6.3}$$

$$B = \begin{bmatrix} 0 \\ 0.9 \\ -1 \end{bmatrix} ; C = \begin{bmatrix} 1 & 0 & 0 \end{bmatrix} \tag{10.6.4}$$

From (10.6.1) and (10.6.2), it can be seen that the relationship between $u(k)$ and $y(k)$ can be expressed as

$$y(k) - 0.5y(k-1) = -u(k+1) + 0.9u(k) \tag{10.6.5}$$

and the corresponding parameter vector is given by

$$\theta = [-\eta_2, -\eta_1, -\eta_0, \beta_1, \beta_0] \tag{10.6.6}$$

Clearly, this is a non-causal system and the preliminary output feedback should be applied to obtain a causal preliminary closed-loop system. Actually, any non-zero feedback gain will make the system causal in this case. The one-step-ahead predictive adaptive control algorithm (Goodwin and Sin, [57]) is used to design the adaptive controller for $v(k)$ and the simulation is carried out as follows:

i) Choose initial preliminary $G(0) = 1$ (this step is only required to enable a simulation of the non-causal system);

ii) Simulate the preliminary closed loop system;

iii) Use $u(k)$ and $y(k)$ to estimate the unknown parameters of the original system;

iv) Calculate the gain $G(k)$ using the estimation results;

v) Evaluate the control signal $v(k)$ by utilizing the one-step-ahead predictive control law and then go to (b))

In the simulation, the reference input $y_r(k)$ (see Fig. 10.4) is a square wave with magnitude $\{\pm 1\}$ and period 200 sampling number. The initial values of the parameter estimates are given by $\hat{\theta}(0) = 0$ and $P(0) = 10^3 I_5$. The simulation results are shown in Figs. 10.5-7. Figure 10.5 shows the parameter estimation results $\hat{\theta}(k)$, Figure 10.6 shows the behaviour of the preliminary feedback gain $G(k)$ and Figure 10.7 displays the response of the closed loop system $y(k)$ together with the reference input $y_r(k)$. From these results it can be seen that the proposed method produces an excellent response after the initial tuning transient and is a good approach for the adaptive control of unknown singular systems.

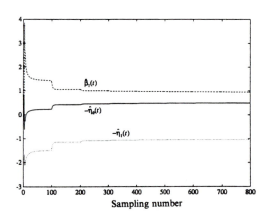

Figure 10.5: Parameter estimation.

10.7 Application to The Control of A Gas Turbine

In this section, we will present the results of the recent application of the above adaptive control strategy for singular system to the control of a simulated gas turbine, which is actually modelled by a 14th order state space equation. The work was recently carried out by Daley and Wang ([40]), where the comparison to standard one-step-ahead adaptive control was also made. Gas turbines are widely used in a variety of power generation

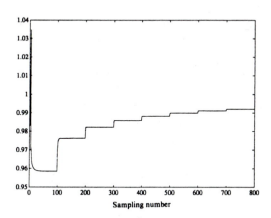

Figure 10.6: Preliminary gain.

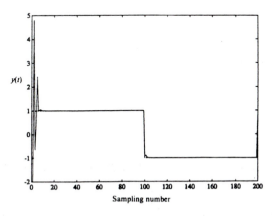

Figure 10.7: Closed-loop system response.

and propulsion applications. The drive for increased efficiency, work ratio and economy is leading to increasingly complex systems with a resulting high demand on the performance of the control system. The turbine dynamics are often complex and vary with operating point and ambient conditions. As a result there has been much recent research into the use of robust and adaptive control for gas turbines.

Strictly speaking, the gas turbine dynamic system is not singular. However, due to its nature of complexity and high dynamic orders, model reduction has to be carried out in order to produce a reasonable simple model for the purpose of controller design. The reduced order model is therefore composed by a set of differential equations and static equations. As a result, a singular system with $N = 0$ (see Eq. (10.2.8)) can be obtained. This fact enables us to use either the standard adaptive control algorithm or singular adaptive control to build up control strategies.

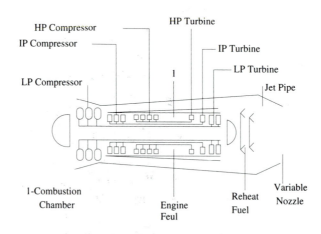

Figure 10.8: Schematic of three-shaft gas turbine.

The original model for the gas turbine shown in Fig. 10.8 was expressed as

$$x(k+1) = Ax(k) + Bu(k) \qquad (10.7.1)$$
$$y(k) = Cx(k) \qquad (10.7.2)$$

where

$$x^T(k) = [N_L, N_l, N_H, P_{2LM}, P_{2l}, P_2, T_3, P_{4H}, P_{4I}, P_{4M}, W_H, W_c, P_5, T_6] \qquad (10.7.3)$$
$$u^T(k) = [W_{FE}, W_{FR}, A_J] \qquad (10.7.4)$$
$$y^T(k) = [W_1, W_2, T_H] \qquad (10.7.5)$$

are the system state vector, the input vector and output vector, with the following notations:

- N_L LP shaft speed;

- N_I IP shaft speed;

- N_H HP shaft speed;

- P_{2LM} LP/IP intercompressor pressure;

- P_{2l} IP/P intercompressor pressure;

- P_2 Combustor pressure;

- T_3 Combustor outlet temperature;

- P_{4H} HP/IP interturbine pressure;

- P_{4I} IP/LP interturbine pressure;

- P_{4M} Post-turbine pressure;

- W_H Hot stream mass flow;

- W_c Cold stream mass flow;

- P_5 Jet pipe pressure;

- T_6 Jet pipe outlet temperature;

- W_{FE} Engine fuel;

- W_{FR} Reheat fuel;

- A_J Nozzle area;

- W_1 Fan mass flow;

- W_2 HP compressor mass flow;

- P_6 Nozzle pressure;

- T_{2LM} LP/IP intercompressor temperature;

- T_{2I} IP/HP intercompressor temperature;

- T_H Thrust

The system, although fourteenth order, is characterized by three dominant eigenvalues. and as shown in [40], can be represented by the reduced order description.

$$x_1(k+1) = A_1 x_1(k) + B_1 u(k) \tag{10.7.6}$$
$$x_2(k) = -B_2 u(k) \tag{10.7.7}$$

where

$$x^T(k) = [x_1(k), x_2(k)] \tag{10.7.8}$$
$$x_1^T = [N_L, N_I, N_H] \tag{10.7.9}$$
$$x_2^T = [P_{2LM}, P_{2I}, P_2, T_3, P_{4H}, P_{4I}, P_{4M}, W_H, W_c, P_5, T_6] \tag{10.7.10}$$

Eqs. (10.7.6) and (10.7.7) can be equivalently expressed by the following singular state space description.

$$Ex(k+1) = Ax(k) + Bu(k) \tag{10.7.11}$$
$$y(k) = Cx(k) \tag{10.7.12}$$

where E is the singular matrix and

$$E = \begin{bmatrix} I & 0 \\ 0 & 0 \end{bmatrix}; A = \begin{bmatrix} A_1 & 0 \\ 0 & I \end{bmatrix} \tag{10.7.13}$$

$$B = \begin{bmatrix} B_1 \\ B_2 \end{bmatrix} \tag{10.7.14}$$

The input-output relation of the reduced order system for parameter estimation purposes can be expressed as

$$u(k) = \eta_3 y(k) + \eta_2 y(k-1) + \eta_1 y(k-2) + \eta_0 y(k-3) + \beta_2 u(k-1) + \beta_1 u(k-2) \tag{10.7.15}$$

and therefore in this case

$$\theta^T = [\eta_3 \quad \eta_2 \quad \eta_1 \quad \eta_0 \quad \beta_2 \quad \beta_1] \in R^{3 \times 18} \tag{10.7.16}$$

The full order system was simulated using Matlab. The initial settings for parameter estimation are

$$P_1(0) = 90I_{18}; P_2(0) = 140I_{18}; P_3(0) = 15I_{18} \tag{10.7.17}$$

$$\hat{\theta}^T(0) = \begin{bmatrix} 5 & 0.2 & 0 & 0 & . & . & . & 0 \\ 0 & 10 & 0.3 & 0 & . & . & . & 0 \\ -0.5 & 0 & 3 & 0 & . & . & . & 0 \end{bmatrix} \in R^{3 \times 18} \tag{10.7.18}$$

The setpoint changes during the course of the simulation is described by

$$y_r(t) = \begin{cases} (0.2, 0, 0)^T; & t \le 5 \\ (1.5, 0, 0)^T; & 5 < t \le 7 \\ (0.2, 0, 0)^T; & 7 < t \le 9 \\ (2.4, 0, 1)^T; & t > 9 \end{cases} \tag{10.7.19}$$

The parameter estimation result for

$$\theta_1 = [\eta_3(1,1) \quad \eta_3(2,1) \quad \eta_3(3,1)] \tag{10.7.20}$$

is shown in Fig. 10.9 and the closed loop system output responses are displayed in Fig. 10.10. It can be seen that the system is stable and desired tracking with respect to $y_r(k)$ has been obtained.

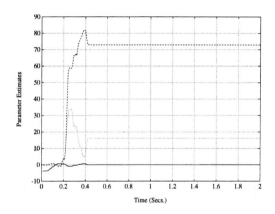

Figure 10.9: Parameter estimation result for singular approach.

Since the reduced system (10.7.6)-(10.7.7) is causal, a standard MIMO one-step-ahead adaptive control [57] can also be employed. For this purpose, an input-output relation of the form

$$\begin{aligned} y(t) &= \alpha_1 y(k-1) + \alpha_2 y(k-2) + \alpha_3 y(k-3) \\ &\quad + \beta_0 u(k) + \beta_1 u(k-1) + \beta_2 u(t-2) \end{aligned} \tag{10.7.21}$$

is used where α_i and β_j are 3-by-3 unknown matrices to be estimated. The control law for this case is

$$\begin{aligned} u(t) &= \hat{\beta}_0^{-1}(y_r(k) - \hat{\alpha}_1 y(k-1) - \hat{\alpha}_2 y(k-2) - \hat{\alpha}_3 y(k-3) - \\ &\quad - \hat{\beta}_1 u(k-1) - \hat{\beta}_2 u(k-2)) \end{aligned} \tag{10.7.22}$$

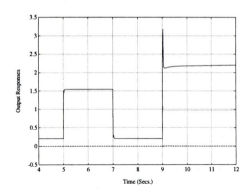

Figure 10.10: System response for singular approach.

The initial settings for parameter estimation are the same as before except that due to the inverse in the control law

$$
\hat{\beta}_0(0) = \begin{bmatrix} 5 & 0.2 & 0 \\ 0 & 10 & 0.3 \\ -0.5 & 0 & 3 \end{bmatrix}
\tag{10.7.23}
$$

which is a well conditioned matrix. The estimates of $\beta_0(1,1), \beta_0(2,1)$ and $\beta_0(3,1)$ are shown in Fig 10.11 and the closed-loop system response is given in Fig. 10.12. It can be seen that the closed loop system is stable but the magnitude of output responses i too large to be acceptable, although parameter estimation is convergent. The reason fo this is because the MIMO one-step-ahead adaptive control needs to calculate the inverse of estimated matrix $\hat{\beta}_0$ at each iteration and good conditioning for this matrix cannot be guaranteed. A comparison of this controller to the standard one-step-ahead controller ha been made by Daley and Wang[40].

10.8   Conclusions

Due to the non-causality of many singular systems, the residual signal which forms the basis of parameter estimation algorithms has to be modified in order that adaptive contro can be applied. This form of the residual enables identification of both causal and non causal unknown singular systems and therefore leads to a novel redesign procedure fo adaptive control algorithms. Instead of identifying the state space form directly, the inpu output form is identified and is then used to construct a preliminary feedback gain, such that the resulting system is input output causal. It has also been shown that, with thi causal relation, any adaptive control algorithm can be applied to singular systems. Severa

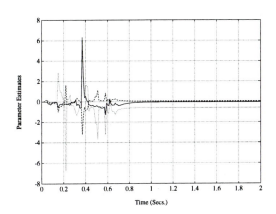

Figure 10.11: Parameter estimation result for standard approach.

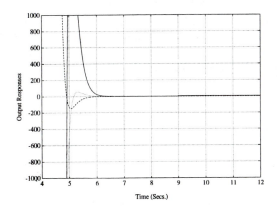

Figure 10.12: System response for standard approach.

simulation examples have been given in order to demonstrate the potential ability of the approach. Indeed, the recent application of the method to the control of a gas turbine has shown that a desired closed loop performance can be obtained by using the algorithm we discussed in this chapter.

Bibliography

[1] Albus, J. S. *A new approach to manipulator control: The Cerebellar Model Artic-ulation Controller (CMAC)*, Trans. ASME J. Dyn. Sys., Mea. & Control, Vol. **63**, No. 3, pp. 220-233, 1975.

[2] An, P. E., W. T. Miller and P. C. Parks, *Design improvements in associative mem-ories for cerebellar model articulation controllers (CMAC)*, Proc. Intl. Conf. on Ar-tificial Neural Networks, Helsinki, North Holland, Vol. **2**, pp. 1207-1210, 1991.

[3] An, P. E., M. Brown, D. J. Mills and C. J. Harris, *High order instantaneous learning algorithms for on-line training*, Proc. IMACS SPRANN '94, pp. 525-529, 1994.

[4] An, P. E., M. Brown and C. J. Harris, *A global gradient noise covariance expression for stationary real Gaussian inputs*, to be published in IEEE Trans. Neural Networks, 1994.

[5] Anderson, B. D. O., and J. B. Moore, *Linear Optimal Control*, Prentice-hall, 1971.

[6] Aslam-mir, S., D. McLean, P. E. An, M. Brown and C. J. Harris, *Active control of large flexible aerospace structures*, SPIE, Orlando, Florida, April, 1994.

[7] Astrom, K. J., *Introduction to Stochastic Control Theory*, Academic Press, New York, 1970.

[8] Astrom, K. J., and B. Wittenmark, *On self-tuning regulators*, Automatica, Vol. **9**, pp. 468-478, 1973.

[9] Astrom, K. J., and B. Wittenmark, *Self-tuning controllers based on pole-zero place-ment*, Proc. IEE, Part D, Vol. **127**, pp. 120-130, 1980.

[10] Astrom, K. J., and T. Hagglund, *Automatic tuning of simple regulators with specifi-cation of phase and gain margins*, Automatica, Vol. **20**, pp. 645-651, 1984.

[11] Astrom, K. J., J.J. Anton and K.E. Arzen, *Expert control*, Automatica, Vol. **22**, pp. 277-286, 1986.

[12] Åström K.J., and B. Wittenmark, *Adaptive Control*, Addison Wesley, Reading, MA, 1989.

[13] Bellman, R., *Introduction to Matrix Analysis*, TATA McGraw-Hill, New Delhi, 1974.

[14] Bossley, K. M., D. J. Mills and M. Brown and C. J. Harris, *Neurofuzzy high dimensional approximation*, in "Symposium on adaptive computing and information", UNICOM 94, 1994.

[15] Brown, M., and C.J. Harris *A nonlinear adaptive controller: a comparison between fuzzy logic control and neuralcontrol*, IMA J. Maths. Contr. and Info., Vol. **8**, pp. 239-265, 1991.

[16] Brown, M., and C.J. Harris *The B-spline Neurocontroller*, In Parallel Processing in Real-time Control, E. Rogers and Y.Li (Eds), Prentice-Hall, 1992.

[17] Brown, M., C. J. Harris and P. C. Parks, *The interpolation capabilities of the binary CMAC*, Neural Networks, Vol.**6** No. 3, pp.429-440, 1993.

[18] Brown, M., P. E. An and C. J. Harris, *A stability analysis of the modified NLMS rules*, submitted to IEEE Trans. Signal Processing, 1994.

[19] Brown, M., and C. J. Harris, *Neurofuzzy Adaptive Modelling and Control*, Prentice Hall, Hemel Hempstead, 1994.

[20] Buckley, P. S., *Automatic control of processes with dead time*, Proc. IFAC World Cong., Moscow, pp. 33-40, 1960.

[21] Canudas de Wit, C., and J. Carrillo, *A modified EW-RLS algorithm for systems with bounded disturbances*, Automatica, Vol. **26**, pp. 599-606, 1990.

[22] Chen, S., and S. A. Billings, *Neural networks for nonlinear dynamic system modelling and identification*, in "Advances in Intelligent Control", Ed C.J. Harris, Taylor & Francis, London, pp. 85-112, 1994.

[23] Chen, H.F., and L.Guo, *Convergence rate of least-square identification and adaptive control for stochastic systems*, Int.J.Control, Vol. **44**, pp. 1459-1476, 1986.

[24] Chen, S., C. F. N. Cowan and P. M. Grant, *Orthogonal least squares learning for radial basis function networks*, IEEE Trans. Neural Networks, Vol. **2**, pp. 302-309, 1991.

[25] Chvatal, V., *Linear Programming*, W. H. Freeman and Company, New York, 1983.

[26] Clarke, D. W., and P. J. Gawthrop, *Self-tuning controller*, Proc. IEE, Part. D Vol. **122**, pp. 929-934, 1975.

[27] Clarke, D. W., and P. J. Gawthrop, *Self-tuning control*, Proc. IEE, Part. D Vol. **126**, pp. 633-640, 1979.

[28] Clarke, D. W., C. Mohtadi and P. S. Tuffs, *Generalized predictive control —- Part I. the basic algorithm*, Automatica, Vol. **22**, no. 2, pp. 137-148, 1987.

[29] Clough, D. E., and S.J. Park, *A novel dead-time adaptive controller*, IFAC Workshop on Adaptive Control of Chemical Processes, Frankfurt, 1985.

[30] Cook, G., and M. G. Price, *Comments on 'A comparison of the Smith predictor and optimal design approaches for systems with delay in the control'*, IEEE Trans. IECI, Vol. **25**, pp. 180-181, 1978.

[31] Craven, P., and G. Wahba, *Smoothing noisy data with spline functions*, Numeric. Math., Vol. **31**, pp. 317-403, 1979.

[32] Cohen, A., and G. A. Coon, *Theoretical consideration for continuous flow systems*, Trans. , ASME, Vol. **75**, pp. 827-833, 1953.

[33] Cohon, J. L., *Multiobjective Programming and Planning*, Academic Press, New York, 1978.

[34] Cox, M. G., *Algorithms for spline curves and surfaces*, NPL Report DITC 166/90, 36 pages, 1990.

[35] Dahleh, M. A., and J. B. Pearson, l^1 *-optimal controllers for MIMO discrete-time systems*, IEEE Trans. Automat. Control, Vol. **32**, pp. 314-322, 1987.

[36] Dai, L., *Observers for singular systems*, IEEE Trans. Automat. Control, Vol. **33**, pp. 187-191, 1988.

[37] Dai, L., *Impulsive modes and causality in singular systems*, Int. J. Control, Vol. **50**, pp. 1267-1281, 1989.

[38] Daley, S., and H. Wang, *On the use of rule based methods for the determination of structure in the identification of process models*, Proc .21th Annual Pittsburgh Conference on Modelling and Simulations, 1990.

[39] Daley, S., and H. Wang, *An adaptive controller for abrupt changing MIMO systems with nonlinear model uncertainties*, Proceedings of 3st Conference on Decision and Control, pp. 259-260, 1992.

[40] Daley, S., and H. Wang, *Adaptive gas turbine control using a singular system approach*, Proceedings of International Conference on Control'94, 1994.

[41] Darouach, M., and M. Boutayeb, *Recursive state and parameter estimation of SISO singular systems*, Proc. IEE, Part. D, Vol. **139**, pp. 204-206, 1992.

[42] Donoghue, J.I., *A Comparison of the Smith predictor and optimal design approaches for systems with delay in the Control*, IEEE Trans. IECI, Vol. **24**, pp. 109-117, 1977.

[43] Donoghue, J.I., *Futher comments on 'A comparison of the Smith predictor and optimal design approaches for systems with delay in the control'*, IEEE Trans. IECI, Vol. **25**, pp. 379-380, 1978.

[44] Douglas, S. C., *Family of normalised LMS algorithms*, IEEE Signal Processing Letters, Vol. **1** No. 3, pp. 49-51, 1994.

[45] Doyle, J. C., K. Glover, P. P. Khargoneker and B. A. Francis, *State space solutions to the standard H^2 and H_∞ control problems*, IEEE Trans. Automat. Contr., Vol. **34**, pp. 831-847, 1989.

[46] Driankov, D., H. Hellendoorn and M. Reinfrank, *An Introduction to Fuzzy Control*, Springer-Verlag, Berlin, 1994.

[47] Egardt, B., *Unification of some discrete-time adaptive control systems*, IEEE Trans. Automat. Contr., Vol. **25**, pp. 693-697, 1980.

[48] Ellacott, S. W., *Aspects of the numerical analysis of neural networks*, Acta Numerica, pp. 145-202, 1994.

[49] Fertik, H. A., *Tuning controllers for noisy process*, ISA Trans., Vol. **14**, pp. 292-304, 1975.

[50] Fogel, E., and Y. F. Huang, *On the value of information in system identification – bounded noise case*, Automatica, Vol. **18**, no. 2, pp. 229-238, 1982.

[51] Franklin, G.F., J.D. Powell and Emami-Naeini . *Feedback Control of Dynamic Systems*, Addison-Wesley Publishing Company, 1988.

[52] Friedman, J. H., *Multivariate adaptive regression splines*, The Annals of Statistics, Vol. **19**, No. 1, pp. 1-141, 1991.

[53] Gawthrop, P. J., and M. T. Nihtila, *Identification of time delay using a polynomial identification method*, Systems Control Letters, Vol. **11**, pp. 267-271, 1985.

[54] Geman, S., E. Bienenstock and R. Doursat, *Neural networks and the bias/variance dilemma*, Neural Computation, Vol. **4**, No. 1, pp. 1-58, 1992.

[55] Girosi, F and T. Poggio, *Networks and the best approximation property*, Bioi. Cybern., Vol. **63,** pp.169-176, 1990.

[56] Goodwin, G., C., K. Cain and A. Ramadge, *Multivatriable discrete-time adaptive control systems*, IEEE Trans. on. Automat. Control., Vol. **25**, 1980.

[57] Goodwin, G. C., and K. S., Sin, *Adaptive Filtering, Prediction and Control*, Prentice-Hall, INC, Englewood Cliffs, New Jersey, 1984.

[58] Guillemin, E.A., *Synthesis of Passive Networks*, New York. John Wiley and Son, Inc. 1957.

[59] Habermayer, H., *Robustness properties of Smith predictor control*, IFAC Symposium on Identification and System Parameter Estimation, Hungary, pp. 1129-1134, July, 1991.

[60] Hammarstrom, L. G., and K.V. Waller, *On optimal control of systems with delay in the control*, IEEE Trans. IECI, Vol. **27**, pp. 301-303, 1980.

[61] Hang, C. C., C. H. Tan and W. P. Chan, *A performance study of control systems with dead time*, IEEE Trans. IECI, Vol. **27**, pp. 234-241, 1980.

[62] Hagglund, T., *New estimation techniques for adaptive control*, Ph.D dissertation, Linkoping University, Sweden, 1984.

[63] Harris, C. J., and S. A. Billings, *Self-tuning and Adaptive Control: Theory and Applications*, London, Peter Peregrinus Ltd, 1981.

[64] Harris, C. J., *Advances in intelligent control*, Taylor and Francis, London, 1994.

[65] Harris, C. J., C. G. Moore and M. Brown, *Intelligent Control: Some Aspects of Fuzzy Logic and Neural Networks*, World Scientific Press, London & Singapore, 1993.

[66] Haykin, S., *Adaptive Filter Theory*, Prentice Hall, Englewood Cliffs, NJ, 2nd Edition, 1991.

[67] Haykin, S., *Neural Networks: A Comprehensive Foundation*, IEEE Press, Macmillan, 1994.

[68] Hochen, R.D., S.V. Salehi and J.E. Marshall, *Time-delay mismatch and the performance of predictor control schemes*, Int. J. control, Vol. **38**, pp. 433-447, 1983.

[69] Isermann, R., *Process fault detection based on modelling and estimation method: a survey*, Automatica, Vol. **20**, pp. 387-404, 1984.

[70] Ivakhnenko A.G., *Polynomial theory of complex systems*, IEEE Trans. on System Man and Cybernetics, Vol. **1**, No. 4, pp. 364-378.

[71] Jacobs, R. A., M. I. Jordan, *Learning piecewise control strategies in a modular neural network*, IEEE Trans. on Sys. Man and Cyb., Vol. **23**, No. 3, pp. 337-345, 1993.

[72] Jones, A.H., and B. Porter, *Expert tuners for PID controllers*, Proc IASTED Conference on Computer-Aided Design and Applications, Paris, 1985.

[73] Jones, A.H., *Design of adaptive digital set point tracking PI controllers incorporating expert tuners for multivariable plants*, IFAC Symposium on Adaptive Control and Signal Processing, Glasgow, U.K. 1989.

[74] Jacobs, R.A., and M. I. Jordan, *Learning piecewise control strategies in a modular neural network*, IEEE Trans System Man and Cybernetics, Vol. **23** No. 3, pp. 337-345, 1993.

[75] Kaczmarz, S. *Angenäherte Auflösung von Systemen Linearer Gleichungen*, Bull. Int. Acad. Pol. Sci. Lett and Cl. Sci. Math. Nat. Ser. A., pp. 355-357, 1937 (A translated version appeared in the *Int. J. Control*, Vol. **57**, No. 6, pp. 1269-1271, 1993).

[76] Kalman, R. E., and R. S. Bucy, *New results in linear filtering and prediction theory*, Trans. ASME, Ser. D., J. Basic Eng., Vol. **83**, pp. 95-107, 1961.

[77] Kalaith, T., *Linear System*, Prentice-Hall, Inc., Englwood Cliffs, 1980.

[78] Kavli, T., *ASMOD - an Algorithm for adaptive spline modelling of observation data*, in "Advances in Intelligent Control", Ed. Harris C. J., Taylor and Francis, London, chapter 6, 1994.

[79] Keyser, R.M.C.D., F.M. D'Hulster and A.R.V. Cauwenberghe, *Predictors for the control of time-delay processes*, Journal A, Vol. **22**, pp. 89-90, 1981.

[80] Hertz, J., A. Krogh and R. G. Palmer, *Introduction to the Theory of Neural Computation*, Addison-Wesley Publishing Company, Redwood City, CA, 1991.

[81] Kreisselmeier, G., and K. S. Narendra, *Stable model reference adaptive control in the presence of bounded disturbances*, IEEE Trans. Automat. Contr., Vol. **27**, pp. 1169-1175, 1982.

[82] Landau, I. D., *A survey of model reference adaptive techniques: theory and application*, Automatica, Vol. **10**, pp. 353-379, 1974.

[83] Landau, I.D., *Adaptive Control: the Model Reference Approach*, Marcel Dekker, New York, 1979.

[84] Lane, S., D. Handelman and J. Gelfand, *Theory and development of higher order CMAC neural networks*, IEEE Cont. Sys Mag., April, pp. 23-30, 1992.

[85] Lawrence, A., and C. J. Harris, *A label drawing intelligent control strategy*, in Neural Network for Modelling and Control, Chapman Hall, 1993.

[86] Lin, W., and B. K. Ghosh, *An adaptive controller for systems with unmeasurable disturbance*, Proceedings of 1992 American Control Conference, pp. 1202-1206, 1992.

[87] Linkens, D. A., and J. Nie, *Unified real time approximate reasoning*, in "Advances in Intelligent Control", Ed Harris C. J., Taylor and Francis, London, chapter 11, 1994.

[88] Liu, G.P., *A simple algorithm of adaptive control for systems with time delay*, The Ptoceedings of IFAC Low Cost Automation 1989, Milan, Italy, pp.71-76, 1989.

[89] Liu, G.P., *Model reference adaptive predictor control for time-varying system with variable time delay*, Journal A, Vol. **30**, pp. 18-24, 1989.

[90] Liu, G. P., *Frequency domain approach for critical systems*, Int. J. Control, Vol. **52**, pp. 1507-1519, 1990.

[91] Liu, G. P., and V. Zakian, *Sup regulators for critical systems*, Control Systems Centre Report no. 731, UMIST, U.K., July 1990.

[92] Liu, G. P., and V. Zakian, *Sup regulators*, The 29th IEEE Conference on Decision and Control, Hawaii, pp. 2145-2146, December 1990.

[93] Liu, G.P., *Adaptive predictor control for slowly time-varying systems with variable time delay*, AMSE Periodicals: Advances in Modelling and Simulation, Vol. **20**, pp.9-21, 1990.

[94] Liu, G. P., *Input space and output performance*, Proceedings of the 10th American Control Conference, Boston, Massachusetts, U.S.A., pp. 1197-1198, 1991.

[95] Liu, G. P., *Mean controllers*, The Proceeding of First European Control Conference, Grenoble, France, Vol. **3**, pp. 2508-2513, 1991.

[96] Liu, G.P., and H. Wang, *An adaptive controller for continuous-time systems with unknown varying time-delay*, Proceedings of the Third IEE International Conference on Control 91, Edinburgh, U.K. pp.1084-1088, 1991.

[97] Liu, G. P., *Theory and Design of Critical Control Systems, PhD Thesis*, Control Systems Centre, University of Manchester Institute of Science and Technology, U.K., Feb. 1992.

[98] Liu, G. P., *Disturbance space and sup regulator in discrete time*, Systems and Control Letters, Vol. **18**, pp. 33-38, 1992.

[99] Liu, G. P., *Self-tuning sup regulators for discrete systems*, Proceedings of the American Control Conference, Chicago, U.S.A., pp. 2270-2274, 1992.

[100] Liu, G. P., *Mean regulators for discrete systems*, Proc. IEE, Part. D, Vol. 139, pp. 67-71, 1992.

[101] Liu, G.P., and L.C. Rao, *Adaptive control with predictor for slowly time-varying systems with variable delay*, International Journal of Control and Computer, Vol. **1**, pp. 6-10, 1992.

[102] Liu, G. P., *Input space and output performance in the frequency domain for critical systems*, IEEE Trans. Automat. Control, Vol. **38**, pp.152-155, 1993.

[103] Liu, G. P., *Rejection of persistent and/or transient disturbances*, Proceedings of the 2nd European Control Conference, pp. 2030-2034, Groningen, The Netherlands, 1993.

[104] Ljung, L., and T. Soderstrom, *Theorey and Practice of Recursive Identification*, Cambridge, Mass.: MIT Press, 1983.

[105] Ljung, L. *System Identification: Theory for the User*, Prentice Hall, Englewood Cliffs, NJ, 1987.

[106] Mamdani, E. H., *Application of fuzzy algorithms for control of simple dynamic plant*, Proc. IEE, Vol. **121**, No. 12, pp. 1585-1588, 1974.

[107] Mason, J. C., and P. C. Parks, *Selection of neural network structures - some approximation theory guidelines*, in "Neural Networks for Control Systems", Eds Warwick K., Irwin G. W., Hunt K. J., Peter Peregrinus, Stevenage, pp. 151-180, 1992.

[108] McCormick, G. P., *Nonlinear Programming, Theory, Algorithm, and Applications*, John Wiley & Sons, New York, 1983.

[109] Meditch, J. S., *Stochastic Optimal Linear Estimation and Control*, McGraw-Hill, New York, 1969.

[110] Middleton, R. H., and Goodwin, G. C., *Digital Control and Estimation, A Unified Approach*, Prentice-Hall, Englewwod Cliffs, N.J., 1993.

[111] Miller, W.T., *Real-time application of neural networks for sensor based control of robots with vision*, IEEE Trans. on Syst. Man and Cybernetics, Vol.**19,** pp. 825-831, 1989.

[112] Millnert, M., *Identification and control of systems subject to abrupt changes*, Ph.D dissertation, Linkoping Univ., Sweden, 1982.

[113] Møller, M., *Efficient Training of Feed-Forward Neural Networks*, PhD Thesis, Computer Science Department, Aarhus University, Denmark, 1993.

[114] Moore, C. G., and C. J. Harris, *Indirect adaptive fuzzy logic control*, in "Advances in Intelligent Control", Ed Harris C. J., Taylor and Francis, London, chapter 12, 1994.

[115] Monopoli, R. V., *Model reference adaptive control with an augmented error signal*, IEEE Trans. Automat. Contr., Vol. **19**, pp. 474-482, 1974.

[116] Narendra, K. S., and L. S. Valavani, *Stable adaptive controller design: direct control*, IEEE Trans. Automat. Contr., Vol. **23**, pp. 570-583, 1978.

[117] Narendra, K. S., Y. H. Lin and L. S. Valavani, *Stable adaptive controller design, part II: proof of stability*, IEEE Trans. Automat. Contr., Vol. **25**, pp. 440-449, 1980.

[118] Narendra, K. S., and R. V. Monopoli, *Application of Adaptive control*, Academic Press, New York. 1980.

[119] Narendra, K. S., and A. M. Annaswamy, *Robust adaptive control in the presence of bounded disturbances*, IEEE Trans. Automat. Control, Vol. **31**, pp. 306-315, 1986.

[120] Narendra, K. S., and K. Parthasarathy, *Identification and control of dynamic systems using neural networks*, IEEE Trans. Neural Networks, Vol. **1**, pp.4-27, 1990.

[121] Ortega, R., and R. Lozano-Leal, *A note on direct adaptive control of systems with bounded disturbances*, Automatica, Vol. **23**, pp. 253-254, 1987.

[122] Ortega, R., and R. Lozano-Leal, *Reformulation of the parameter identification problem for systems with bounded disturbances*, Automatica, Vol. **23**, pp. 247-251, 1987.

[123] Palmor, Z., *Stability properties of Smith dead-time compensator controllers*, Int. J. Control, Vol. **32**, pp. 937-949, 1980.

[124] Parlos, A. G., A. F. Henry, F. C. Schweppe, L. A. Gould and D. D. Lanning, *Nonlinear multivariable control of unclear power plants based on the unknown-but-bounded disturbance model*, IEEE Trans. Automat Control, Vol. **33**, pp. 130-137, 1988.

[125] Parks, P. C, *Liaypunov redesign of model reference adaptive control systems*, IEEE trans. Automat. Control, Vol. **11**, pp. 362-367, 1966.

[126] Parks, P. C., and J. Militzer, *Improved allocation of weights for associative memory storage in learning control systems*, Proc. 1st IFAC Symp. on Design Methods for Control Systems, Zurich, Pergamon Press, pp. 777-782, 1991.

[127] Peterson, B., and K. S. Narendra, *Bounded error adaptive control*, IEEE Trans. Automat. Control, Vol. **27**, pp. 1161-1168, 1982.

[128] Polycarpou, M. M., and P.A.Ioannou, *Leaning and convergence analysis of neural-type structured networks*, IEEE Tans. on. Neural Networks, Vol.**3,** pp. 39-50, 1992.

[129] Post, E., *Formal reduction of general combinational problems*, Am Journal of Math, Vol. **65**, pp. 197-268, 1943.

[130] Procyk, T. J., and E. H. Mamdani, *A Linguistic Self-Organising Process Controller*, Automatica, Vol. **15**, pp. 15-30, 1979.

[131] Rumelhart, D. E., and J. L. McClelland (Eds), *Parallel Distributed Processing: Explorations in the Microstructure of Cognition*, MIT Press, Cambridge, MA, 1986.

[132] Sage, A. P., and C. C. White III, *Optimum Systems Control*, Prentice-Hall, Inc Englewood Cliffs, N.J., 1977.

[133] Samson, C., *Stability analysis of adaptively controlled systems subject to bounded disturbances*, Automatica, Vol. **19**, pp. 81-86, 1983.

[134] Sanner, R.M and J.E.Slotine, *Gaussian Networks for direct adaptive control*, IEEE Trans. on Neural Networks, 1991.

[135] Sanger, T.D. *Neural network learning control of robot manipulators using gradually increasing task difficulty*, IEEE Trans. Robotics and Automation, Vol. **10**, No. 3, pp. 323-333, 1994.

[136] Sargent, R. A. E., *Expert systems for process control rooms*, Measurement and Control, Vol. **19**, pp. 239-244, 1986.

[137] Sarpturk S, Y.Istefanopulos and O.Kaynak, *On the stability of discrete-time sliding mode control systems*, IEEE Trans. on. Automat. Control., Vol. **32,** pp. 930-932, 1987.

[138] Shewchuk, J. R., *An Introduction to the Conjugate Gradient Method Without the Agonizing Pain*, Technical Report CMU-CS-94-125, School of Computer Science, Carnegie Mellon University, Pittsburgh, PA, 1994.

[139] Shin, Y., and J. Ghosh, *Ridge polynomial networks*, to appear in IEEE Trans. Neural Networks, 1994.

[140] Sinha, N.K., and B. Kuszta, *Modeling and Identification of Dynamic System*, Van Nostrand Reinhold Press, 1983.

[141] Smith, O.J.M., *Closer control of loops with dead time*, Chemical Engineering Progress, Vol. **53**, pp.217-219, 1957.

[142] Smith, O.J.M., *A controller to overcome dead time*, ISA J., pp. 28-33, Feb. 1959.

[143] Slotine,J.E. *Sliding controller design for nonlinear systems*, Int.J.Control., Vol.**40,** pp. 421-428, 1984.

[144] Souza, C.E.D., G.C. Goodwin D.Q. Mayne and M. Palaniswami, *An adaptive control algorithm for linear systems having unknown time delay*, Automatica, Vol. **24**, pp. 327-341, 1988.

[145] Sugeno, M., and M. Nishida, *Fuzzy control of model car*, Fuzzy Sets and Systems, Vol. **16**, pp. 103-113, 1985.

[146] Tresp, V., J. Hollatz and S. Ahmad, *Network structuring and training using rule-based knowledge*, in "Advances in Neural Information Processing Systems 5", Eds Giles C. L., Hanson S. J., Cowan J. D., Morgan Kaufman, San Mateo, CA, 1993.

[147] Verghese, G.C, B.C.Levy and T. Kailath, *A generalized state-space for singular systems*, IEEE Trans. Automat. Control, Vol. **26**, pp. 811-831, 1981.

[148] Vidyasagar, M., *Optimal rejection of persistent bounded disturbances*, IEEE Trans. Automat. Control, Vol. **31**, pp. 527-534, 1986.

[149] Wang, H., and A. H. Jones, *A rule based identifier for unknown systems*, Proc. IEE, Part. D, Vol. **138**, pp. 500-506, 1991.

[150] Wang, H., Y.Q. Liu and D.H. You *Application of a nonlinear self-tuning controller for regulating the speed of a hydraulic turbine*, Trans. ASME. J. Dynamic Syst. Measurement and Control. Vol.**113,** pp. 541-544, 1991.

[151] Wang, H., and S. Daley, *Controller tuning via rule based identification*, Trans. Measu. Contr., Vol. **15**, pp. 38-45, 1993.

[152] Wang, H., and A. P. Wang, *Towards stable neural network controllers*, Proceedings of International Control Conference on Control 94, Vol. **1**. , 1994.

[153] Wang, H., and C. J. Harris, *Identification of a family of unknown nonlinear systems via fuzzy logic*, Internal Report, TR03/92, 1992.

[154] Wang, H., M. Brown and C. J. Harris, *Modelling and control nonlinear, operating point dependent systems via associative memory networks*, Int. J. Dynamic and Control, Kluwer Academic Press, to appear, 1995.

[155] Wang, H., M. Brwon and C. J. Harris, *Fault detection for a class of unknown nonlinear systems via associative memory networks*, Proc. IEE, Part. D, Vol. **208**, pp. 101-108, 1994.

[156] Wang, H., and C. J. Harris, *Fuzzy logic based robust servomechanical controller design for nonlinear unknown systems*, Proceedings of 9th International Conference on Systems Engineering, pp. 435-439, 1993.

[157] Wang, H., and S. Daley, *Identification and adaptive control for SISO singular systems*, Int. J. Syst. Science, Vol. **24**, pp. 1791-1801, 1993.

[158] Wang, L. X., *Stable adaptive fuzzy control of nonlinear systems*, IEEE Trans. Neural Networks, Vol. **1**, pp. 146-155, 1993.

[159] Wang, L. X., *Adaptive Fuzzy Systems and Control: Design and Stability Analysis*, Prentice Hall, Englewood Cliffs, NJ, 1994.

[160] Watanabe, K., Y. Ishiyama and M. Ito, *Modified Smith predictor control for multivariable systems with delays and unmeasurable step disturbances*, Int. J. Control, Vol. **37**, pp.959-973, 1983.

[161] Watanabe, K., *Adaptive Estimation and Control, Partitioning Approach*, Prentice Hall, 1992.

[162] Wellstead, P. E., M. J. Edmunds, D. Prager and P. Zanker, *Self-tuning pole/zero assignment regulators*, Int. J. Control, Vol. **30**, pp. 1-26, 1979.

[163] Wellstead, P. E., D. Prager and P. Zanker, *Pole assignment self-tuning regulator*, Proc. IEE, Part. D, Vol. **126**, pp. 781-787, 1979.

[164] Wellstead, P. E., and D. Prager, *Multivariable pole assignment self tuning regulators*, IEE Part. D, Vol. **128**, 1981.

[165] Werntges, H. W., *Partitions of unity improve neural function approximation*, Proc. IEEE Int. Conf. Neural Networks, San Francisco, CA, Vol. **2**, pp. 914-918, 1993.

[166] White, D. A., and D. A. Sofge, *The Handbook of Intelligent Control*, Van Nostrand Reinhold Inc., NY, 1992.

[167] Whildborne, J.F., and G.P. Liu, *Critical Control Systems: Theory, Design and Applications*, Research Studies Press Limited., 1993.

[168] Widrow, B., R.G.Winter and R.A. Baxter, *Layered neural nets for pattern recognition*, IEEE Trans. Acoust.. Speech. Signal Processing, Vol.**36**, No. 7, pp. 1109-1118, 1988.

[169] Widrow, B., and M. A. Lehr, *30 years of adaptive neural networks: perceptron, madaline, and backpropagation*, Proc. of IEEE, Vol. **78**, pp.1415-1441, 1990.

[170] Willsky, A. S., and H. L. Jones, *A generalized likelihood ratio approach to the detection and estimation of jumps in linear systems*, IEEE Trans on Automat Control, Vol. **21**, pp. 108-112, 1976.

[171] Wonham, W. M., *On the separation theorem of stochastic control*, SIAM J. Control, Vol. **6**, pp. 312-326, 1968.

[172] Wu, Q.H., and B.W.Hogg, *Adaptive controller for a turbogenerator system*, IEE Part D, Vol. **135,** pp. 274-279, 1988.

[173] Xie,X., and R.Evans, *Discrete-time adaptive control for deterministic time-varying systems*, Automatica, Vol. **20,** pp. 309-319, 1984.

[174] Yip, E.L., and R.F.Sincovec, *Solvability, controllability, and observability of continuous descriptor systems*, IEEE Trans. Automat. Control, Vol. **26**, pp.702-707, 1981.

[175] Youla, D. C., H. A. Jabr, and J. J. Bongiorno, *Modern Wiener-Hopf design of optimal controllers-Part 2: the multivariable case*, IEEE Trans. on Automat. Control, Vol. **21**, pp. 319-338, 1976.

[176] Zadeh, L. A., *Fuzzy Sets*, Information and Control, Vol. **8**, pp. 338-353, 1965.

[177] Zadeh, L. A., *Outline of a new approach to the analysis of complex systems and decision processes*, IEEE Trans. Sys. Man and Cyb., Vol.**3**, No. 1, pp. 28-44, 1973.

[178] Zhang, B. S., and J. M. Edmunds, *Self-organizing fuzzy logic controller*, IEE Proc. D, Vol. **139**, No. 5, pp. 460-464, 1992.

[179] Zakian, V., *A performance criterion*, Int. J. Control, Vol. **43**, pp. 921-931, 1986.

[180] Zakian, V., *On performance criteria*, Int. J. Control, Vol. **43**, pp. 1089-1092, 1986.

[181] Zakian, V., *Input space and output performance*, Int. J. Control, Vol. **46**, pp. 185-191, 1987.

[182] Zakian, V., *Critical systems and tolerable inputs*, Int. J. Control, Vol. **49**, pp. 1285-1289, 1989.

[183] Ziegler, J. G., and N. B. Nichols, *Optimum settings for automatic controllers*, Trans ASME, Vol. **64**, pp.759-768, 1942.

Index